SCIENCE AND FAITH

Two Faces of the Same Fact

SCIENCE AND FAITH

Two Faces of the Same Fact

Edited by
Mustafa Mencüketin

BLUE DOME

Copyright © 2012 by Blue Dome Press
15 14 13 12   1 2 3 4

𝕱ountain Books

All rights reserved. No part of this book may be reproduced or transmitted in any form or by
any means, electronic or mechanical, including photocopying, recording or by any information
storage and retrieval system without permission in writing from the Publisher.

Published by Blue Dome Press
345 Clifton Ave., Clifton,
NJ, 07011, USA

www.bluedomepress.com

Library of Congress Cataloging-in-Publication Data Available

ISBN: 978-1-935295-23-5

Printed by
Çağlayan A.Ş., Izmir - Turkey

CONTENTS

Editor's Preface

When we extend the timeline of modernity back to Cartesian dualism of early 17th century, we can see that the idea of religious faith and scientific thought as two separate realms of knowledge has been a problematic one in modern times.

With the birth and growth of postmodernity in 20th and 21st centuries that has been deconstructing and solubilizing almost all the primordial values of humanity, we observe that hypothetical dualism has been regarded as the sine qua non of man's advance for the better: People—explicitly or implicitly—are constrained to chose "either religious faith or scientific thought", but only one of them.

This is of no surprise as Ali Ünal infers on further pages that *"the separation of science and religion and assigning to each a different realm of competence or relevance is responsible for religion being seen as a set of myths and dogmas—blind beliefs— and science remaining in the darkness of materialism."*

That be pointed out, this modest collection of articles previously published in the *Fountain Magazine* handles the issue of scientific thought and religious faith with utter precision and endeavor to present a clear stance, one as far from subjectivity as possible.

If existence, the universe, deeds, and phenomena are viewed through a prism of unity, it becomes as clear as daylight that religious faith and scientific thought are two fountains springing from the same unique source.

To exemplify this notion, Zeki Sarıtoprak analyses "the Ultimate Cause" under the light of the premise that *"the laws of nature proceed from God's Wisdom"* and that *"behind every cause and effect is the Cause of causes (Musabbib al-asbab)*. Ali Sebetci offers to discuss the conceptualization of the sciences in an unprecedented way:

"Most of us assume that humankind is continually improving in its knowledge of existence and thinking so as to understand reality as the ages pass. ... The modern conceptualization of the sciences is subservient to the idea that reality can only be examined by its highly specific and restricted experimental method. Thus, the modern conceptualization implies, due to this method, that theories, doctrines, and principles concerning non-observable realities cannot be scientific. A natural consequence of this way of thinking is to be skeptical about metaphysical realities in the belief that the absolute reality is the physical one. From this point of view, therefore, neither religion nor metaphysics are accepted as scientific disciplines."

Hamza Aydın invites readers to notice and activate four indispensable connections in the universe that *"individuals, groups, and societies to have a healthy life as well as to maintain their health at all levels depend on."* Osman Çakmak calls attention to the magic of macro realms specifying our need to *"travel toward the horizon of spirit and develop an all-new scientific language which approaches physics and metaphysics together... the physical and observable universe which is the domain of research for modern physics is an ornamented curtain veiling the world of the unseen"*, while Olgun Hasgül seeks an answer to the question *"when does human life begin?"* On the same vein, Salih Adem attempts to prism *"the universe in the light of modern physics"*; Dr. Şenol Ersin philosophizes on mystical affiliations between the sub-atomic world and creation. Fatih Kocabaş accompanies you on a specific *"quest to solve the mystery of life,"* as Alphonse Dougan handles science and religion in the same analytic circle *"between friction and harmony."*

As Fethullah Gülen underlines, *"Humankind has suffered from depraved, ambitious souls who believe only that 'might is right.' This will continue until humankind builds a world on the foundation of science and faith."*

We hope that upon reading this book, you gain new insights into the notion of faith and science, not as separate opposing entities as mistakenly popularized, but rather as one combined entity stemming from the same source. Now enjoy the insightful journey into the mysteries of science and faith!

Misunderstandings about Science and Technology

■ Fethullah Gülen

Today, we are as exposed to great calamities and dangers as we are to great possibilities and hopes. The nineteenth and twentieth centuries have been two of the most productive and richest centuries for both good and evil. Had humanity made proper use of the opportunities of this period, we might have changed the world into a paradise.

The general atmosphere and conditions surrounding us promise great happiness so long as science and technology with all their facilities and fruits are devoted to the service of humanity. Nevertheless, up to now we have not been able to make proper use of these extensive possibilities. The happiness of humanity has been delayed. There is doubt and despair because nobody has taken on the responsibility for guiding humanity to eternal bliss. This will probably continue until those who have authority and competence come to acknowledge their true responsibility.

In the past, cities and villages used to live in an isolation that was unpleasant, disorderly and devoid of spirituality. They were frequently visited by privation and pestilence and were permeated by immorality. Moreover, they merely observed what was going on with indifference, believing life to be no more than this. Today, it is obviously impossible to live with out-of-date conceptions disconnected from reality. 'The continuance of the old state being impossible, we must either follow the new order or be annihilated' We will either reshape our world as required by science, or we shall destroy the earth and ourselves.

Some believe that science will reduce humans to machines, and humankind to a swarm of ants, running the world in accordance with a mechanistic and deterministic worldview. This is not true at all! Science has existed throughout human history, and everything is in the end connected with it. A world without science has nothing to give to humans.

It is, however, true that in some of our cities people have been reduced, human feelings have diminished, and the virtue of health in mind and body have been wiped out. But it is an injustice to ascribe all

this only to science and technology. The fault lies rather with scientists who avoid their responsibilities. Many traumatic events would probably not have happened if scientists had acquired an awareness of their social responsibility and had performed what was expected of them.

Science means comprehending what things and events tell us, and what the Divine Laws prevailing over the universe reveal to us. It means striving to understand the purpose of the Creator. Humans, who have been created to rule over all things, need to observe, read, discern and learn about what is around us. Then, we have to seek to exert our influence over events and subjecting them to ourselves. At this point, by the decree of the Sublime Creator, everything will submit to humans, as we shall submit to God.

Science, with all its branches such as physics, chemistry, astronomy, and medicine, is at the service of humanity, and every day brings new gains which may also be gifts of hope. There is no reason for us to be afraid of science. The danger does not lie with science itself and the founding of a world in accordance with it, but rather with ignorance, and the irresponsibility of scientists.

We acknowledge that some intentions based on knowledge may sometimes give bad results, but it is certain that ignorance and disorganization always give bad results. For this reason, instead of being opposed to the products of science and technology, it is necessary to use them so as to bring us happiness. Herein lies the essence of the greatest problem of humankind. It is simply not possible to go back in time before the space age and remove the thought of making atomic or hydrogen bombs from the minds of people.

Given that science might be a deadly weapon in the hands of an irresponsible minority, we should nonetheless adopt science with its products to found a civilization where we will be able to secure our happiness in this world and the next. It is vain to curse the machine and factories, because machines will continue to run and factories to operate. So too science and its products will not cease to be harmful to humankind until the seekers of truth assume the direction of things and events.

We must not fear science nor the technology it enables, for such a fear paralyses us. Instead, we must be fearful of the hands using it. Sci-

ence in the hands of an irresponsible minority is something disastrous which could on its own transform the world into hell. Having come to know only after the destruction of Hiroshima and Nagasaki that a monster of cruelty had exploited his studies of the atom, Einstein apologized in tears to his Japanese counterparts, but by then it was too late!

That calamity was not the first and won't be the last of its kind. Seas have become sewage and poison, rivers have become canals of filth, and the atmosphere has become a fog of pollution, all due to the barbarian minority, and this will continue to be so...

Humankind has never suffered harm from a weapon in the hands of angels. Whatever they have suffered has all come from depraved, ambitious souls who believe only that "might is right." This will continue until humankind builds a world on the foundation of science and faith. It is our hope that human beings will come to comprehend the nature of the world we live in before it is too late.

Quantum Entanglement: Illusion or Reality?

■ Süleyman Çandaroğlu

S cience has always influenced philosophy. When scientific thought changes, even in slow and apparently trivial ways, social thought also goes through profound changes. For example, by the end of the nineteenth century, the classical sciences had been developed so rigorously that they became dominant in the life of the individual and society. The effect of this domination can also be seen in the last two or three centuries in environmental issues such as the destruction of flora and fauna and industrial pollution. The classical approach to the way nature works was mechanical, deterministic, and materialistic. Science was reductionist, denying the understanding of complexity which is nowadays known to be one of the most important challenges science faces. This reductionist approach proceeds as though understanding the working principle of a component part of an object or event makes it feasible to discover the working principles or future trajectories of "the whole" by using classical science. This view of life is overly simplistic. Applying these reductionist principles to social life and human thought causes disaster.

The quantum description of the universe is very different from the classically observed one, or from our perceptions in everyday life. This new way of looking at nature has many consequences, both philosophical and practical. The modern technological development of the second half of the last century may be a very good example of the consequences of the discovery of the quantum world. Now we have gadgets from cellular phones to long-lasting batteries, from engineered drugs to space missions, from pocket-size computers to nanotechnology, a wide range of end-products of the quantum world. These changes and innovations will certainly continue, and quantum sciences will eventually affect the way we look at life.

One of the most dramatic potential changes in thought may arise from the discovery of the quantum entanglement of particles. Quantum en-

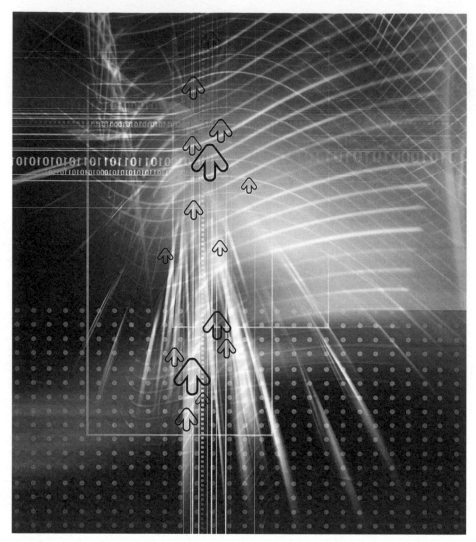

tanglement can be described as non-classical correlations of different particles. It is very different from the classical description and can be explained by using the following analogy. Imagine an author writes a book of one hundred pages on the most precious arts or explaining very important facts about the universe. To make it more interesting or more realistic, he gives different pages of the book to different people and asks them to read and understand the rules written in the book. Each person has access to only one page of the book. If we assume the information on the pages follows the classical formula, each person has a hundredth of the total information written in the book; and if we let them communicate with each other, they can in principle reconstruct the book. However, the situation is very strange in the quantum world. If the information in the book is written using the entanglement principle of the quantum world, then no one person has any definite idea about the partial information on his page. It is as if the pages are empty. All the information is contained in the correlations and relationships between each part, not in each disparate part. So each individual has no idea what the book is about, if he has only one page.

Einstein vs. Bohr

To understand this strange feature of quantum entanglement we should review the historical development of the concept. One of the earliest objections came from Einstein, who was one of the developers of quantum theory. Although he explained the photoelectric effect by introducing the concept of quantization of light, he did not believe in some of its consequences. Mainly, he was not sure about the completeness of quantum theory because it contradicted both common sense and the theory of relativity. The famous 1927 Solvey Conference was a turning point for debates between Einstein and Niels Bohr, who was also one of the developers of quantum theory and specifically of the Copenhagen interpretations of this theory.

Einstein tried to illustrate this incompleteness by proposing different *Gedanken* (thought) experiments. Each of these questions was answered rigorously by Bohr, but he never managed to convince Einstein of the completeness of the theory. The last of these *Gedanken* experiments was related to our concept, quantum entanglement. It is called the EPR paradox and takes its name from the authors of the famous paper, *"Can a quantum mechanical description of physical reality be considered complete?"* by Einstein, Podolsky and Rosen in 1935.

Mainly, the paper was about faster-than-light communication between physically separated objects, two particles. If two particles are generated from a source affected by the existence of a conservation law, such as the conservation of energy or linear or angular momentum, the conserved property is carried by the particles independent of their separation. If the conserved quantity is observed by measuring one of

the particles, the other particle arranges itself according to the result of this measurement independent of the distance between particles. According to Bohr, this arrangement happens instantaneously at the time of measurement, which conflicts with Einstein's theory of special relativity, which says nothing can travel faster than light. Apparently, the knowledge of the result of the first measurement is carried somehow to the second particle. Bohr's reply is now called the Copenhagen interpretation of quantum mechanics. He takes this property as a postulate of quantum mechanics by saying that the state of the particles includes all information about them. After he published this explanation, Einstein never replied again.

If we look more closely at the proposed experiment, we can deduce that information is not transferred faster than light, because although the measurement result of the second particle is decided by the first measurement, this information is hidden from the second particle. The result of the second measurement makes sense only if the result of the first measurement reaches the second one. Otherwise, the second measurement can be described as a random outcome. Now we can illustrate this with our previous book analogy. Here our book has only two pages, each given to two people. If they only look at their own page there is no information, which means that measurement results are random.

However, if the two people work together and share their measurement results, then the initial information can be reconstructed.

Coins

Einstein's point of view can be described with the following example. Imagine we have two coins with the usual heads and tails. Let us assume that there is a conservation law deduced from everyday experiments stating that if we flip these two coins we always have two opposite results; that is, if we get tails from the one that we measured, the other one is heads, and vice versa. In the real world, these coins can be identified as electrons, photons or atoms. Heads/tails corresponds to the spin components for electrons, polarization directions for photons or ground/excited states for atoms. Now, imagine these two coins are separated by a large distance.

Einstein says that as soon as separation occurs the result of the flip is decided, but this result is hidden from us. One can measure or learn it by performing a measurement or looking at each coin. Moreover, looking at only one coin is enough to determine the result of the other coin, since the results are correlated. Conjecturing that the result is determined at the time of measurement is against the causality principle of the theory of relativity, which says that cause and effect cannot be simultaneous. But I am of the opinion that reality is closer to what Bohr described. That is, the result of the measurement is decided at the time of measurement, not at the time of separation. Before the measurement, each coin shows both heads and tails at the same time. The information, taken at the point of measurement when one of the coins is measured, is transferred faster than light; in other words, at infinite speed.

The nature of each coin is also very strange before the measurement because it includes both sides at the same time with equal probability, but a classical coin has only one side at a time, either heads or tails. Here the classical coin means the flipped or measured coin. This property of the quantum world is called parallelism. As in the famous case of Schrodinger's cat, sometimes two extreme situations can happen at the same time. Schrodinger's cat is a very special cat which is dead and alive at the same time, like a quantum coin. However, when one measures such a cat, that is, observes the cat, its nature collapses to one of the known situations, either a dead cat or a live cat. This measuring process happens systematically due to interactions with its surroundings, and is called decoherence.

Although the quantum world is very strange and different from the classical world, it encapsulates more reality than we experience in our everyday life. In the near future, we can expect that ways of looking at the world will be different from the present mechanical, deterministic, and materialistic view because of the unexpected outcomes of the quantum world. If you know how to look, you can already feel this change.

Nature: A Book to Be Contemplated

■ Dr. Zeki Sarıtoprak

Nature is a wonderfully designed masterpiece of art. As humanity is in continuous contact with nature, our relationship with it has always been significant. What does nature mean to us? Although we have a strong relationship with nature, the quality of this relationship is a matter of discussion. Some naturalist scientists accept nature as so independent that they consider it self-governing and self-creating; other scholars consider every aspect of nature as dependent upon the Master Architect of the universe.

As a result of the former view, our modern approach toward nature is often aggressive and arrogant. Until recently, people in the West viewed nature as something to be conquered and thus dominated. Humanity's resulting struggle to conquer and tame nature was inevitable. In the 1960s, when British scholars managed to measure the top of Mount Everest, they declared that they had conquered Everest, the object of their efforts. This is a clear indication of our modern approach to nature: domination instead of harmony[1].

The world's great religions have significantly different approaches to nature. This article analyzes the Islamic approach by focusing on the Qur'anic understanding of nature as a book to be contemplated.

Islam and Nature

Islam views nature as a book similar to the Qur'an, through which God reveals Himself. Although the Qur'an does not explicitly declare nature to be the book of God, it contains many references to nature as parts of the limitless words of God. One verse reads: "Though the sea became ink for the words of my Lord, verily the sea would be used up before the words of my Lord were exhausted, even though we brought the like thereof to help" (18:109). Some Islamic scholars interpret this as indicat-

---

1   Cited in Ian G. Barbour, *Religion in an Age of Science: The Gifford Lectures* 1989-1991 (USA: Harper San Francisco Press, 1990), 1:245.

ing that the sea itself is a sort of word of God. Therefore, if the sea were doubled it still could not provide enough ink to exhaust God's words, because the ink in question is His word. Therefore, one can view every part of nature as a sentence or word in the book of nature, the author of which is God.

Accordingly, studying nature means studying the book of God. Those who believe in the Qur'an are expected to read this book in order to understand God, contemplate His creation, and maintain a dialogue with Him. Humanity is endowed with the ability to think, and thus can see God's Beautiful Names reflected throughout the natural world. This idea is indicated in:

> In the creation of the heavens and Earth and the succession of day and night are signs for people of wisdom; those who remember God standing and sitting and lying on their sides, and who ponder over the creation of the heavens and Earth, exclaiming: Our Lord, You have not created this in vain (3:119).

God's Beautiful Names[2]

By contemplating nature, we inevitably realize the meanings and reflections of God's Beautiful Names. As we are the only conscious creatures in the physical world, only we are required to reflect upon nature. The Islamic tradition states that God has 99 Names, some of which are reflected in the natural world.

For example, the sun with all its glory, as well as the moon and the stars, reflect the Name An-Nur (The Light). Earth, with all its living creatures, reflects Al-Hayy (The Living One). The whole universe is the result and reflection of Al-Khaliq (The Creator). All love in the universe, from the love of human mothers for their children to the love of animal mothers for their young (in other words all human and animal beings' inherent capacity to love) reflect Al-Wadud (The Loving One). At the end of his *Divine Comedy*, Dante (1265-1321) indicates that God's love is the love that moves the sun and other stars. This idea was also held by Al-Jami, a famous Sufi. Many Muslim theologians, especially Sufis, consider the sun's movement to be a result of the ecstatic love it feels for God.

Another example is the sustenance of all creatures, which is the product of Ar-Razzaq (The Sustainer). The capacity to see or understand reflects Al-Basir (The Seeing One). The uniqueness of all living things and the va-

---

2   For a complete list of God's Beautiful Names, see Ian S. Markham, *A World Religions Reader* (Oxford: Blackwell Publishers, 2000). He also gives an excellent account of Al-Ghazali's *The Ninety-nine Beautiful Names of God*, translated by David. B. Burrell and Nazih Daher.

riety of shapes reflects Al-Musawwir (The Fashioner). All things of beauty reflect Al-Jamil (The Most Beautiful One). These examples present nature's positive aspects and how they reflect God's Names. However, its negative aspects (e.g., natural disasters and death) also reflect some of His Names.

The Names and Nature

The lofty reality of the relationship between God's Names and nature perhaps can be understood as similar to the relationship between an object and a mirror. The sun's manifestation and reflection appear in all small fragments of glass and droplets of water. They reflect the sun but are not the sun itself. Therefore, all creatures and the natural world reflect God but are not God. This element distinguishes Islam from pantheism, which sees nature as God.

Given that nature reflects God's Names, it was created for a purpose. From the planetary to the atomic level, nature contains no randomness or chance of accidental creation. Everything is consciously created with a measure, which engenders an inimitable art. Even the smallest items of creation are crafted artfully and in painstaking detail. Such verses as: "It is He [God] who creates and then disposes, Who measures and then

guides" (87:3), and "We have created everything with a measure" (54:49) support the idea that everything in the universe is measured and guided by God, the All-Powerful and All-Wise, Who has directly designed and created all things. Islamic scholars assert that God must design and create even the smallest thing, for if He did not there would be chaos in the world. Accordingly, the universe is not a completely closed system of laws of cause and effect.

The Qur'an calls the regularity of natural phenomena and the relationship between natural causes and their effects *"sunnat Allah"* (God's way of acting). If one recovers from an illness by taking medicine, the Islamic understanding indicates that medicine is the apparent cause. However, the real Healer is the one who gives medicine its ability to heal. There is a subtle relationship between healing and medicine, because medicine and the process of healing itself both reflect As-Shafi (The Healer). Therefore, the search for medicine is a search for As-Shafi, something to be attained and not denied.

Likewise, you must follow the natural process if you want to produce an apple: plant a seed, wait for the tree to grow, water the tree, and see that it receives enough of the sun's heat and light. In other words, oxygen, hydrogen, nitrogen, and carbon combine in different ways to compose all plants, and God uses this process to work His Will.

Cause and Effect

The process of creation is the result of this relationship of cause and effect, out of which determinism and the famous Newtonian Mechanism are derived. The relationship between cause and effect is sometimes very strong, sometimes very subtle, and sometimes indiscernible. For example, we can calculate astronomically at which time the sun will rise 10 years from now, but we cannot calculate or know if there will be rain this time next year. The scientific explanation of rain cannot determine when it will rain until the physical signs are present. Likewise, we may know life's characteristics but not what ultimately makes something live. The Qur'anic understanding contrasts with Newton's model, indicating that God is continually active in the universe's creation: "every day He exercises universal power" (55:29).

The laws of nature proceed from God's Wisdom. Behind every action and reaction is the Cause of causes (*Musabbib al-asbab*), **the ultimate Cause**. In fact, the nature of creation indicates that the causes we see in our daily life cannot create the effects we witness. Natural causes—oxygen, hydrogen, nitrogen, and carbon—are basically orderless, formless, blind, and unconscious, and therefore cannot create anything that is so consciously well-designed. For example, the elements we receive from food are consciously distributed to the appropriate bodily organs and, in fact, neither we nor the elements are aware of this wonderfully organized distribution. Said Nursi, a contemporary Muslim scholar, argues that:

> If we deny the All Powerful One who is the Cause of causes, then we should accept that in every particular working in your eye there would have to be an eye which could see every limb and part of your body as well as the entire universe with which you are connected[3].

As the natural world's causes depend upon the ultimate Cause, they can be suspended. The regularity of natural phenomena depends upon the command of The Most Powerful One. He can suspend all natural laws, for He is the One Who enacted them. Examples of this are the miracles that He allows the Prophets to perform. Fire's nature is to burn, but it does not burn when God does not will it to, as in: "We said: O fire, be calm and peaceful toward Abraham" (21:69). The miracles of Jesus, Moses, and Muhammad are other examples. To fully explore miracles and their implications is beyond the scope of this paper.

When God wills the creation of something, He commands: "Be! (*kun*)" and it is. Natural laws obey His commands and therefore do not act independently. This is demonstrated when His Prophets are protected through a suspension of natural law. Muslim scholars interpret apparent causes as veils that hide God's Power or Will. His Wisdom necessitates a veil between us and His Majesty.

In this world, *dar al-hikmah* (the Domain of Wisdom), His Wisdom dominates. Thus the relationship of cause and effect proceeds from His Wisdom. God can create a full-grown person in one day, but His Wisdom

---

3   Nursi, Said, *The Treatise on Nature, The Flashes Collection* (the 23rd Flash on nature) (Istanbul: Sozler Publication, 1995), 232-54. Nursi argues that the naturalists' way is impossible, whereas the Qur'an's way is necessary. He defends this idea through parables and comparisons. In the introduction of his treatise he writes: "This treatise puts naturalistic atheism to death with no chance of resurrection, and totally shatters the foundation stones of unbelief."

necessitates that a person be created in a way that allows him or her to have various stages of life. By working through a process in nature, God teaches people that they must follow a similar process in their deeds. Although God's Power is reflected in the universe's creation and in the natural world's process of cause and effect, Its real reflection is in the Hereafter (*dar al-qudrah* [The Domain of Power]), which contains no cause or effect, but only God's power directly.

Conclusion

The Qur'anic understanding of nature encourages us to acknowledge our ecological responsibility to the environment, because nature, being a book of God, deserves our highest respect. The Qur'an promotes love for nature, because nature is a result of the love of God as well as a guide leading to the discovery of ultimate reality. Through contemplating nature and being more aware of the reflections of God's beautiful Names, believers see His signs and experience His presence in their daily life. They also will observe, through nature, His art and the lesson He gives to humanity through His creation.

The Islamic Conceptualization of the Sciences

■ Ali Sebetci

W e believe it is not emphasized today as much as it should be that traditional conceptualizations of sciences, whether Islamic or not, were very different from the way modern sciences are currently conceptualized. Most of us assume that humankind is continually improving in its knowledge of existence and reflecting so as to understand reality as the ages pass. This may be true, especially when the detailed information of physical reality gathered and the level of technology reached with the help of the modern scientific method are considered. However, it is usually missed by modern minds that there is a crucial difference between the modern and traditional conceptualization of sciences which veils the significance of traditional sciences in the contemplation of the Absolute Reality. In this article, we would like to try to cast some light on this difference from the Islamic point of view, which is the last manifestation of all the traditional ones.

Starting with the following classification of the sciences laid out in Shams al-Din Muhammed al-Amuli's *Nafa'is al-Funun* (Precious Elements of the Sciences), a treatise written during the fifteenth century, may be helpful to reveal this point [4]:

The theoretical aspects of what we call positive sciences today (physics, chemistry, biology, medicine, geology, astronomy, cosmology, etc.) were all called natural philosophy not only by Islamic scholars but also by the scientists, philosophers, and theologians of other traditions up to the Age of Enlightenment, and that type of science was considered a lower science in the hierarchy of existence. The intermediate science was mathematics, which included music, while metaphysics occupied the place of the highest science. Not only philosophy, ethics, economics, and politics but also intellectual and transmitted religious studies are covered by the Islamic conceptualization of sciences. The classification

---

4   Nasr, Seyyed Hossein. *Islamic Science, An Illustrated Study*, World of Islam Festival Publishing Company Ltd, 1976.

The Islamic Conceptualization of the Sciences

of sciences was always considered to be important in Islamic tradition and many Muslim scholars have provided different classifications[5]. The essential point here is that all the Islamic classifications of sciences, as in this specific one, were based on the Islamic doctrine that there is a hierarchy of existence (and therefore knowledge) from Unity to multiplicity; from the Divine Names and Qualities to each creature on a particular level of existence; and from the Meta-cosmic Reality to the cosmos consisting of the seven heavens and the earth, which is nothing but the manifestation of the Meta-cosmic Reality. Therefore, all sciences, in the Islamic conceptualization, were means of gaining this kind of knowledge.

In contrast, the modern conceptualization of the sciences is subservient to the idea that reality can only be examined by its highly specific and restricted experimental method. Thus, the modern conceptualization implies that theories, doctrines, and principles concerning non-observable realities cannot be scientific. A natural consequence of this way of thinking is scepticism about metaphysical realities, for example in the belief that the absolute reality is the physical one. From this point of view, therefore, neither religion nor metaphysics are accepted as scientific disciplines. Although there are some departments such as politics, economics, and ethics under the umbrella of humanities and arts, the traditional occult sciences such as astrology, alchemy, magic, and the interpretation of dreams cannot find a place in modern universities, in spite of the fact that their derivatives are still alive in both traditional and modern societies. At this point it may be illustrative to remind the reader of the following verses of the Qur'an: *"Thus, did We establish Joseph in the land (Egypt), that We would impart to him knowledge and understanding of the inner meaning of events, including dreams"* (Yusuf 12:21). Also,

> Solomon never disbelieved. Rather, the satans disbelieved, teaching people sorcery and the (distorted form of the) knowledge that was sent down on Harut and Marut, the two angels in Babylon. And they (these two angels charged with teaching people some occult sciences such as breaking a spell and protection against sorcery) never taught them to anyone without first warning, "We are a trial, so do not disbelieve." (Al-Baqarah 2:102)

---

5   Bakar, Osman. *Classification of Knowledge in Islam*, Islamic Texts Society, 1999.

This can be interpreted as evidence for the existence of such sciences, although the Qur'an itself condemns them:

> And (yet) they learned from them (the two angels) that by which they might divide a man and his wife. But (though they wrongly attributed creative power to sorcery, in fact) they could not harm anyone thereby save by the leave of God. And they learned what would harm them, not what would profit them. Assuredly, they knew well that he who bought it (in exchange for God's Book) will have no share in the Hereafter. How evil was that for which they sold themselves; and if only they had known. (Al-Baqarah 2:102)

In the Islamic conceptualization of sciences, the natural sciences possessed a place in the hierarchy of knowledge so that there were no contradictions between them and religion. The whole worldview was religious, and this perspective provided sufficient space for the cultivation of positive sciences in the guidance of metaphysical principles from a higher level of knowledge. The history of Islamic sciences is full of striking examples of success from the modern scientific perspective in cosmology, geology, mineralogy, botany, zoology, mathematics, astronomy, physics, medicine, pharmacology, agriculture, and irrigation, any of which could be the topic of a future article. This is why, as people learn ever more natural sciences, they become more devout believers. That is why most of the greatest Muslim scientists were also distinguished philosophers and theologians. They are those who the Qur'an identifies in the phrase, "Of all His servants, only those possessed of true knowledge stand in awe of God" (Al-Fatir 35:28).

It is because of the modern conceptualization of sciences that religion and science are separated from each other and have become strangers, in the assumption that they have their own realms to speak about[6]. The roots of this way of thinking go back to Descartes (1596–1650):

> Descartes, who appeared with the thesis that 'metaphysics cannot be a science,' said that knowledge can only be obtained by the investigation of measurable and divisible things, and he limited the question of science to only matter; since that time the followers of Cartesian philosophy have always talked in a similar way.[7]

---

6   Edis, Taner. *An Illusion of Harmony: Science and Religion in Islam*, Prometheus Books, 2007.
7   Gülen, Fethullah. www.herkul.org, Kırık Testi, September 12, 2005.

Yet, both religion and science try to find answers to the questions of who we are, what the meaning of life is, where we come from, and where we are going. We cannot imagine any technological advancement which satisfies these eternal concerns of humanity and reflects its relation to eternity.

By definition, the modern sciences have bound themselves to the existence of a finite and relative reality. Their methodology is very effective and there is nothing wrong with it unless it is used to abuse natural sources or to corrupt the natural order. Those sciences may even provide some clues about the higher level of existence, as in the case of quantum mechanics, which has destroyed the deterministic worldview of classical (Newtonian) physics. However, to our mind, there is a problem with the modern conceptualization of sciences, which neither accepts any notion of the hierarchy of existence nor has any means of studying and investigating those levels of reality.

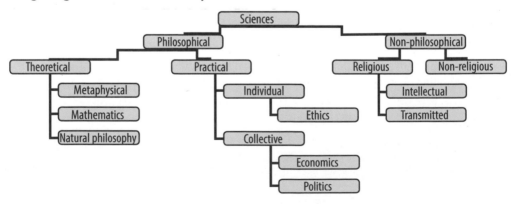

Mathematics and the Universe

■ Ali Kemal Ünver

People have very different attitudes to mathematics. While some love it, some find it very difficult and some even hate it. Even though it is true that mathematics is built on an axiomatic foundation, a strong case can be made that the ultimate foundation of mathematics is its beauty. Richard Feynman, an American physicist known for expanding the theory of quantum electrodynamics, says, "To those who do not know mathematics it is difficult to get across a real feeling as to the beauty, the deepest beauty, of nature... If you want to learn about nature, to appreciate nature, it is necessary to understand the language that she speaks in."

Educators who see the beauty at the center of mathematics and can make their students see it that way are more likely to be able to get their students' attention and teach them more effectively. This method facilitates the study of sophisticated mathematics and puts mathematics in its proper place, so as better to understand its practical value for human beings. To see the beauty and the pleasure in mathematics can change the negative attitudes of some students and help mathematics teachers.

Often it seems that we pursue the study of mathematics from either a structural or an applications point of view. From a structural point of view, we insist on building up all of the tools one may need in a sequential, logical order, because an educator will need the students to know all of the smaller pieces before they can build up to larger ideas. An analogy for this would be if we forced somebody to study all of the nails, screws, bolts, and tools to build a house before we let them even see the plans for the house. This is one of the main reasons that most people who study mathematics at school think that it is a pointless exercise in playing with formulas that have no significance in real life. For these people, mathematics might be helpful after graduation only in keeping track of their checkbooks. Some students think that they can calculate whatever they need using computers; but sometimes this can be ineffective if students

33

do not understand the logic behind the problem. In such situations students will not understand the significance of the answers the computer gives, and will not be able to detect basic errors of input.

The applications point of view leads to creating "word problems" that appear to be about the real world, but which are highly artificial. It also leads to focusing on higher levels only when dealing with applications. Hence, for instance, in calculus we spend a lot of time plodding through various physical applications, without letting students see the bigger picture. Or we spend time in liberal arts mathematics talking about things such as modeling and linear programming, which yield useful applications, but are generally tedious and do not give most students much appreciation for mathematics. If we see the beauty at the center of mathematics, as well as introducing ideas that may be of practical use, we can bring students to a much bigger perspective on mathematics at a much earlier stage in their mathematical development.

Educators can include fun topics in courses that they teach. For instance, they can encourage students to discover the amazing number patterns in nature, such as the Fibonacci sequence in pine cone spirals, pineapples, and cauliflowers, in which the number of pieces increases in the following manner: 0, 1, 1, 2, 3, 5, 8, 13, 21, 34, 55, 89 ... (add the last two numbers to get the next). Another example is the "golden ratio." In mathematics and the arts, two quantities are in the golden ratio if the ratio between the sum of those quantities and the larger one is the same as

the ratio between the larger one and the smaller. The golden ratio is a mathematical constant, approximately equal to 1.6180339887 and recent research shows that people think that the shapes and figures in this ratio are more interesting and aesthetically pleasing to the human eye.

Sometimes, even a small algebra trick can miraculously bring students to love mathematics and be more focused, and then more interested in the deeper aspects later on. Take a look at this symmetry:

1 x 1 = 1
11 x 11 = 121
111 x 111 = 12321
1111 x 1111 = 1234321
11111 x 11111 = 123454321
111111 x 111111 = 12345654321
1111111 x 1111111 = 1234567654321
11111111 x 11111111 = 123456787654321
111111111 x 111111111 = 12345678987654321.

Here are a few more examples showing the beauty of mathematics visually with numbers:

1 x 9 + 2 = 11
12 x 9 + 3 = 111
123 x 9 + 4 = 1111
1234 x 9 + 5 = 11111
12345 x 9 + 6 = 111111
123456 x 9 + 7 = 1111111
1234567 x 9 + 8 = 11111111
12345678 x 9 + 9 = 111111111
123456789 x 9 +10 = 1111111111

$9 \times 9 + 7 = 88$

$98 \times 9 + 6 = 888$

$987 \times 9 + 5 = 8888$

$9876 \times 9 + 4 = 88888$

$98765 \times 9 + 3 = 888888$

$987654 \times 9 + 2 = 8888888$

$9876543 \times 9 + 1 = 88888888$

$98765432 \times 9 + 0 = 888888888.$

Students' minds can be broadened by seeing the surprising differences that arise when we move to non-Euclidean geometry. Fractals in nature, for example snow crystals and the Mandelbrot set, which is a set of points in the complex plane whose boundary forms a fractal, can be introduced to give a background and context to the lesson. These beautiful images can trigger in the students deep and thought-provoking ideas, fostering curiosity, creativity, and an open mind.

All these examples and others like them can inspire people to see the beauty of mathematics and give them a better understanding and a sense of the expanse of mathematics. While discovering the more beautiful aspects of mathematics, students are much less likely to hate mathematics and may develop a much greater appreciation for the creation of the universe.

Note:

* The golden ratio can be derived by the quadratic formula, by starting with the first number as 1, then solving for the 2nd number x, where the ratio $[x+1]/x = x/1$ or (multiplying by x) yields: $x+1 = x2$, or a quadratic equation: $x2-x-1=0$. Then, by the quadratic formula, for positive $x = [-b + sqrt(b2-4ac)]/2a$ with $a=1, b=-1, c=-1$, the solution for x is: $[-(-1) + sqrt([-1]2 -4*1*-1)]/2*1$ or $[1 + sqrt(5)]/2$. See the second reference for details.

References

1. http://users.forthnet.gr/ath/kimon/

2. Green, Thomas M. "The Pentagram and the Golden Ratio." http://www.contracosta.cc.ca.us/math/pentagram.htm.

Connection, Always and Everywhere

■ Hamza Aydın

> *"... There is a stable order in the world as well as a well-established connection, constant norms and fundamental laws. In this sense, the world is analogous to a clock or a well-designed machine. Every single wheel, every single screw, every single nail not only has a role in the machine and an impact in its final benefit but also positive consequences for all living beings, especially for humans[8]."*

There is a strong parallel between the general principles observed in a healthy social structure in a society and the necessary conditions that enable cells to make up a healthy tissue. The laws prevalent in the universe present amazing parallels since they derive from the same divine source. In order to have healthy development in societies, it is necessary to have healthy connections, reciprocal understanding, and proper communication between people. Similarly, the health of a cell, which can be considered as a micro-society, also depends on the continual use of complex signal connection networks and maintenance of connections with its environment. In order to maintain a harmonious and healthy cell, there must be dense information transfer (through chemical molecules) at all levels with neighboring cells. Thanks to the connection networks that begin in the membranes, cells can recognize warnings coming to them and produce a response.

Through such information networks, a continuous connection is established in living organisms, starting from their lowest level mechanisms (i.e. molecules in cells and organelles) to their highest level mechanisms (i.e. organs, systems, organisms, populations, ecosystems), in order to ensure a healthy and harmonious process of development, reproduction, differentiation and aging. Organized as a tissue (micro-society), one of the most amazing features of cells is the way they behave in accord with the society they live in, rather than acting as individual-

---

8   Nursi, Said. *Signs of Miraculousness*, Seven Heavens, p. 186.

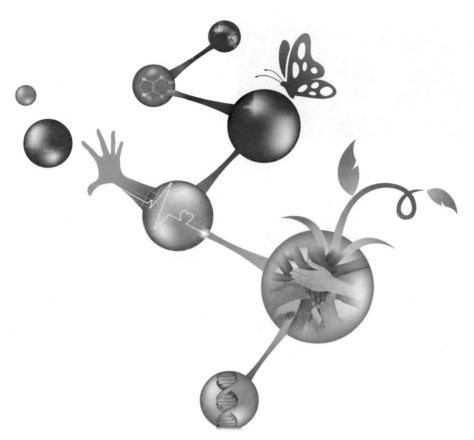

istic beings. Cells behave in a way that makes micro-societies possible. In cytology (the study of cells), this feature is known as "contact inhibition" (i.e. maintenance of healthy and harmonious operation and development based on connection), and its damage may lead to cancer. Every cell is sensitively programmed to receive all signals, coming from inside and outside, and to manage the proper responses to them. Here, the question arises whether a lack or disorder of connection is caused by pathological conditions in cells or whether the emergence of pathological conditions is caused by connection problems. It is generally assumed that molecular changes in the cell occur first (i.e. mutation) and then this mutation leads to abnormalities and connection problems at the level of tissue, organ, and even organism. Damage to communication or connection, and disorders in this system, are a very important reason for the appearance of some pathological conditions, such as an abnormal in-

crease of cells, cancer and death. Thus, diagnosis and treatment of many diseases today depends on knowledge of how biological communication and connection are harmoniously established.

Cells are designed to control their behavior through special signal molecules that can stimulate responses in other cells. For instance, with signal molecules cells can establish colonies or biofilms. Moreover, plant cells, through channels known as plasmodesmata, maintain their connection with neighboring cells and trade certain materials. In a state of illness, the density (among cells) of signal molecules that are transferred in the plasmodesmata changes.

Hormones, reproduction factors and neurotransmitters are charged with ensuring transportation and communication at different levels. Cell and tissue elasticity and their adaptability are increased by this diversity of signal operations. Different methods are used to transmit signals into the cell depending upon the characteristics of the signals. For example, hydrophobic (water-avoiding) molecules like steroid hormones pass directly through the cell membrane and connect to their receptors in the cell. The receptors that are responsible for decoding genes stimulate the decoding of related genes. If a problem (mutation) occurs in the molecules that are responsible for transportation and communication, the transportation and communication breaks down and the cancer process is triggered. The existence of continuously reproducing cells at inappropriate times and places is an important symptom of cancer initiation. For example, in colon and rectum cancer, if a mutation occurs on the Ras protein, which is one of the signal proteins that takes the "reproduce!" message from the cell membrane, the cutting off of the GTP molecule, which is responsible for turning the signal molecule on and off, is blocked; and since the molecule stays permanently active the "reproduce!" signal is continuously sent inside the cell. Thus, the benefits of the medicines that are used in cancer treatment show their effects by hindering signals on the cell membrane level, and this is observed in the non-existence of this mutation on Ras proteins. If the mutation happens the medicines men-

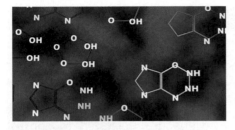

tioned above cannot be effective. Today, the presence of a mutation can be determined by a biopsy taken from the patient; thus, it has become possible to choose the type of therapy that will be most helpful to that person.

Communication inside the cell can be carried out through ion channels, for example sodium, potassium, and calcium. These channels through membranes behave selectively for every different ion. For example, while voltage-gated channels open and close according to electric charge ligand, (key)-gated (receptor) channels let ions transfer when ligands are tied up. Sodium and potassium ions and the molecular channels through which these two pass are responsible for organizing the changes that effect the communication of nerves in the membrane potential. The calcium channel, on the other hand, plays an important role in muscle contraction, and biological events like the formation and deformation of bone.

In recent years, much evidence has shown that the proteins (matrix) that fill the vacancies among cells are the main actor in the general control of communication between cells and in the integration of signals coming from their surroundings. It has been pointed out that CCN proteins, which are one of the adaptors, as well as proteins with multiple modules that are responsible for the connection between cell membranes and matrix proteins, have a regulatory role at different levels in the control of signal transfers in ion channels, cell differentiation, adherence of cells to each other, cell collapse, programmed cell-death, cartilage formation and the synthesis of new veins. The multi-dimensional and dynamic communication and connections mentioned above also apply to people. When individuals develop a healthy connection between their inner world and other people, a healthy society emerges. Every individual is granted these potential connection points, which make the

existence of an individual possible. The development of a healthy person depends upon activating these connections, organizing them dynamically and keeping them active.

If we place human beings at the center of creation, the first connection that the individuals should make between their Creator and their ego (*nafs*) is called worship. The ego is both a help and a hindrance to the construction of this connection. The second connection, which is between individuals and their friends, is ensured by good morals, good conduct, and personal virtue. The construction of a healthy social life depends on how many people have good morals, good conduct and virtuous character in a society. If virtue is not accompanied by knowledge, it is highly unlikely that knowledge will raise an individual to a standard of human perfection.

The third essential, the individual's healthy connection with their surroundings, is established through the "ecological dimension of ego," which is sensitive to the external world. It is very difficult for people whose dimension of ego, which is sensitive to ecological problems and the environment, has not developed to keep the environment clean and take precautions against pollution. The fourth connection that individuals should establish is the connection with their internal world (heart-consciousness, transcendental ego, real self). "O Man! Know thyself!" and, "The one who knows himself knows his God," are expressions pointing out the importance of this connection.

For individuals, groups, and societies to have a healthy life levels depends on activating these four connections; in other words, establishing coordination and harmony among them and then maintaining this state. When one ignores any of these connections, or when the coordination between these connections is defective, troubles and illnesses at various levels emerge. Hence, the links that a healthy individual establishes may include those in civil society organizations, and through these further networked and reciprocal communications. Like our cells, which maintain their connection with their environment and neighboring cells via the "contact inhibition" mechanism so that we feel healthy, for people to become healthy at the personal and societal level, individuals should actively join civil society organizations and service-centered communities that can enable activation of these four connections and ensure harmony and coordination among them. The Islamic scholar, Bediüzzaman Said Nursi, paid special attention to connection in the letters he wrote to his students (*The Kastamonu Letters*) and highlighted it as a point of progress that should be reached:

> Since in today's world, saving people's belief for the sake of God is a very important mission that is above everything; since quantity is unimportant compared to quality; since transient and changing political worlds are trivial compared to the everlasting, constant, stable services in the name of God—they should not be even compared; they can never be objectives; therefore, we should be satisfied with valuable positions that are granted in the circle of the *Risale-i Nur*. Instead of just good intentions towards other people or seeing them at high worldly positions, we need to have extreme loyalty and steadfastness, and utmost connection and sincerity. We should have progress on these points.

Action and Coincidence

■ Nizamettin Yıldız

I t is not easy for people living today to believe that every object, every law and every incident in the universe is planned in a very detailed way. However, it is a fact that there is a seen and unseen algebraic reality to everything moving in the universe. This situation amazes the distinguished scholars roaming on the fringes of science.

Physicists study the biggest and smallest physical measures, the strongest and weakest forces, and they have calculated the ratio between some of them, several times determining this ratio to be close to 1040. For example, it has been proven that the "strong nuclear force," which keeps protons and neutrons in the atomic nucleus together, is 1040 times stronger than the force of gravity. Although some consider this situation a result of coincidence, there are also some people who have shown the courage to question the accuracy of this result and open it to discussion. Since then, obtaining the same number several times has inevitably led to the belief that the number was determined and calculated before. This situation resembles the situation of a chess player who hears somebody telling him the opponent's next move, or the situation of a composer who hears a melody from upstairs that perfectly fits his lyrics while he is trying to write the melody for his lyrics. Is it not amazing when somebody predicts what will be, especially at a time when you least expect such a vision? Questions emerge: Have all actions and incidents in the universe and the values corresponding to them been determined in advance? Is there a certain logic behind the behavior of materials without intellect or consciousness?

Perhaps when Comte de Buffon (1707–1788) started to do research into probability calculations concerning falling matches and needles three centuries ago, he did not think that he would make such an astonishing discovery. According to his calculations, the probability of dropped nee-

dles hitting a pair of parallel lines that are drawn a certain distance apart is proportional to the number pi (π) (Figure 1–2). When the distance between the parallel lines are drawn the length of a match, this probability becomes exactly $2/\pi$. This was a theoretical result that was calculated on paper, but trying it out would raise interesting results. In other words, when a certain number of tests were conducted, it could be expected that a number of matches equal to the number that was found through the theoretical calculations would hit the lines; and, indeed, that was what happened. In addition, mathematicians found another way to calculate pi by using this method, since the ratio of the number of matches or needles that hits the lines to the total number of matches or needles should give a pi-proportioned number. In 1901, Mario Lazzarini threw one needle 3,408 times and, as a result, got the ratio 355/133 or 3.1415929; the difference between this value and the real value is only 0.0000003.

The experimental proof of this fact is not difficult. When the experiment, which has been conducted thousands of times, was first tried by a group of mathematicians dropping 3,000 needles, the number of the needles that touched the lines was close to 1,900, which was the desired result, and the result was amazingly found to be proportionate to pi.

$M/N = 2c/\pi a$ (Figure 3)

Although the real relation is like the equation given in Figure 3, when the length of the needles and the distance between the lines are equalized, the desired ratio becomes $2/\pi$. What does this mean? Does the number pi, which was created with the universe and which we meet in different fields, play a role in showing the manifestation of the Majestic Will

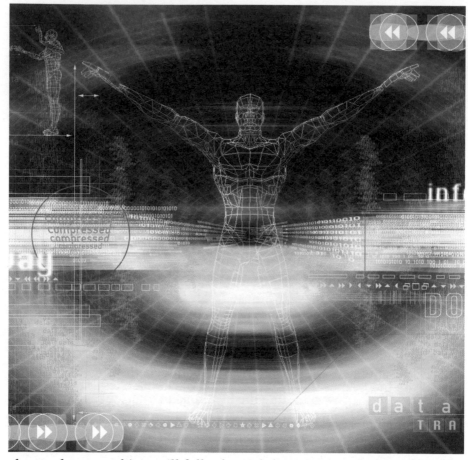

about where an object will fall—through having a result that cannot be explained by coincidence? Is falling not an ordinary incident?

While this reality makes even falling an extraordinary incident, it opens a perspective on understanding the reality behind the verse: "... it was not you (O Messenger) who threw but God threw[9]," which was revealed about the Battle of Badr in the Qur'an. Actually, it is impossible even for a leaf to fall without the knowledge and the calculation of Our Lord, who is closer to us than our jugular vein.

Let us think about a group of creatures that lives with different physical laws in a different universe. Assume that they live on a flat, circular world (Figure 4) and their steps get longer when they come close to the

---

9   Al-Anfal 8:17.

center. For this kind of creature, the shortest distance
between two points is not
a straight line as it is for us
(A–B). Since their steps get
longer as they get closer to
the center, they travel close
to the center. Yet, since they
make the way a little longer
in this way, the shortest distance would be an oblique
line that takes these two
variables into account and
that passes by partially ap-
proaching the center. So,

what would we think if we saw these creatures walking in this way all the
time? Or if we knew that the creatures acting in this way were inanimate
beings? In these circumstances, we might wonder whether these beings
are very intelligent or whether the One who knows and sees everything,
and is present in every place at every time, directs them[10].

For a soccer forward finding the best time to attack when he is facing
the goal keeper, or for a tennis player choosing the best timing and position to hit the ball, a fine calculation is required. In tennis, the player
sometimes approaches very close to the net to meet the ball. In this way,
the player gains great advantage since, by his or her positioning, the
player reduces the area into which the ball can fall to the minimum, and
increases his or her own chances of returning the ball (Figure 5). Nevertheless, since the ball reaches the player faster and harder, there is also
a higher probability of the player's failing to return the ball. Therefore,
advancing right up to the net may not always be advantageous. Thus, the
best position for the player may lie at any point between the net and the
baseline when area and speed variables are considered. Similarly, the
most advantageous point for the goal keeper lies between the attacking
forward and the goal line, at a point which depends upon the variables
of the speed of the ball and the area.

10 "And He knows whatever is on land and in the sea; and not a leaf falls but He knows it" (Al-An'am
    6:59).

Naturally, we do not find tennis and soccer players' positioning as described above strange since we expect them, as reasonable people, to play in this way. Yet, how would we interpret and explain it if we saw inanimate things acting in the same way? There is a phenomenon that applies this logic consciously but itself does not have consciousness. A phenomenon that astonishes people: light.

Let us think of a rectangular racetrack with points A, B, C, D. While the shortest distance for a horse that will run from one end of this racetrack to the other is the AC diagonal when the ground is homogenous, there will be a reroute if the ground is not homogenous. Assume the length of the racetrack is 80 meters and its width is 60 meters. Half of it is grass and the other half is sand (Figure 6). Also, assume the horse's speed on the sand is half of its speed on the grass (it runs 10 meters in 1 unit of time on the grass). Under these circumstances, the horse will run the AC diagonal in 15 units of time while it will run AEC route in 14.5 units of time. The ABC route is much longer. Thus, any route which passes between these two routes will be shorter than these two. The point O, which the shortest route (AOC) passes, will be between the points M and E. When we look carefully, we understand that this is the route light follows as it enters environments of different densities; for example, the

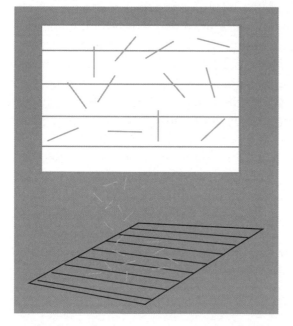

route it follows when passing from air to water (see a spoon's broken image in a water-filled glass). As you see, light finds that specific mysterious point and follows that particular path. In other words, while entering different environments of varying densities, it finds the shortest way and follows it in an amazing way.

Figure 3

$$\frac{M}{N} = \frac{2c}{\pi a}$$

All these detailed calculations and functioning show that even inanimate beings and atoms are in the hands of a Majestic Will.

Every atom contains two truthful testimonies to the Necessarily Existent Being's Existence and Unity. Despite being powerless and insentient, the atom bears decisive witness to the Necessarily Existent Being's Existence by carrying out important duties and functions as though it were conscious. It also testifies to the Unity of the same Being, Who owns all material and immaterial dominions, by conforming to the universal order in general, and to the rules of each place it enters in particular. It settles in every place as if it were its homeland. All of this shows that the One Who owns the atom owns all the places it enters. By carrying out very heavy duties incompatible with its size and weakness, the atom shows that it acts at the command and in the name of the One with absolute power.[11]

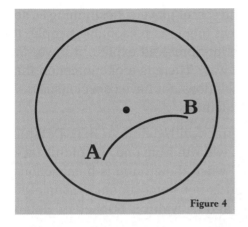

Figure 4

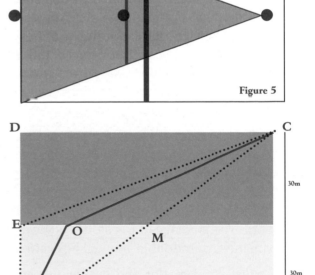

Figure 5

Figure 6

---

11   Nursi, Said. The Thirtieth Word, Second Point, *Risale-i Nur Collection*.

The Holy Qur'an and Dirac's Theory of Pairs

■ Sultan Bashir Mahmood & Robin Yasin Qusab[12]

P aul Dirac (1902–1984), one of Newton's successors to the Lucasian Chair of Mathematics at Cambridge, was one of the greatest physicists of the twentieth century, and his work is one of the monuments of modern physical theory (See Abdus Salam 1966). This article highlights his great work on the possible existence of all matter in pairs, and the revelation of the same theory in the Holy Qur'an, thirteen hundred years before Dirac's research.

Dirac's Discovery

Dirac had the honor of being the first scientist in history to demonstrate the principle that all particles in the universe must exist in pairs, that for each particle there must exist a corresponding anti-particle of exactly the same mass but with an opposite electrical charge. For all matter in the universe, in other words, there must exist an equal amount of anti-matter. Thus the existence of a proton must imply the possible existence of an anti-proton; if a hydrogen atom exists, there must also exist an atom of anti-hydrogen, perhaps in some distant corner of the universe.

Equal amounts of matter and anti-matter must have been produced in the first moments of the Big Bang. In our present universe, however, a particle will not be found co-existing peacefully side by side with its corresponding anti-particle. To understand this, we could recall the old European myth of the doppelganger, the perfect double of the unlucky hero. The message of the stories is always to avoid the doppelganger: if you meet your double you will be destroyed.

Just so, if matter and anti-matter meet they annihilate each other, their energy and momentum dissolving into heat and light. According to Dr. Abdus Salam of the International Centre for Theoretical Physics, "In Dirac's language, anti-matter is 'minus matter'; matter and anti-matter

---

12   Dr. Sultan Bashir Mahmood is the founder of the Holy Qur'an Research Foundation in Islamabad. Mr. Robin Yasin Qusab is a graduate of Cambridge University and a freelance journalist.

just cannot co-exist in the same part of the universe without the ever-impending catastrophe of annihilation; and indeed some astronomers do believe that just this type of annihilation of galaxies and anti-galaxies is taking place at the sites of powerful X-ray sources in the heavens" (See Abdus Salam 1966). Dirac predicted the existence of anti-matter in 1934, and the discoveries of specific anti-particles in the following years confirmed his prediction.

The Fish Problem

According to an apocryphal story, the germ of Dirac's breakthrough materialized from a mathematical brainteaser seemingly unrelated to theoretical physics. During a meeting of the Cambridge Undergraduate Mathematical Society, the Archimedians, the following problem was presented. Those who like a challenge can try working it out for themselves. The answer is provided below for those whose brains become tangled by figures.

After a long day, three fishermen have caught a good amount of fish. They are about to set sail for home when a storm suddenly builds up. Under raging skies, they decide to seek shelter on a nearby island. They unload their catch and set a fire before falling asleep. A few hours later,

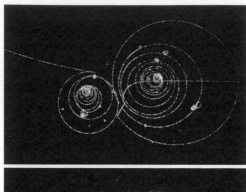

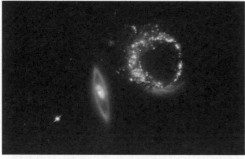

one fisherman wakes up to find that the weather has settled enough for a safe return to be possible. Not wanting to disturb his friends, he divides the haul into three equal parts. There is one fish remaining, and this the fisherman throws back into the sea. He then leaves with his share.

A little later, the second fisherman awakes, also with the desire to get back home. Unaware that one of his friends has already left, he too divides the catch into three equal parts. Again there is one fish remaining, and again this is thrown back into the sea. The fisherman rows off with his portion. Finally, the third fisherman arises and goes through the same process, dividing the remaining fish into three parts, finding one fish remaining, which he then throws back into the sea. He leaves the island with his portion.

And now for the problem: What is the minimum number of fish in the original catch? To put it another way, what is the smallest number which can be subdivided three times, leaving one fish outstanding each time?

Dirac's Answer

Dirac thought for a while before arriving at the answer: minus two. If you divide minus two fish three times, each third will contain minus one fish, with plus one fish outstanding. Each time a fisherman throws the plus one fish into the sea, and rows off with the minus one fish. Expressed in an equation, it looks like this: $-2 = -1-1-1+1$.

Minus one fish is obviously no use to anyone, and this is certainly a brainteaser. It is important only in that it gave birth to the notion of antiparticles in Dirac's mind and can help us to understand the mathematical principle behind his discovery. It led him to the conclusion that all particles exist in opposing pairs and allowed him to see the symmetry of matter and anti-matter.

The Holy Qur'an and Dirac's Theory

As we have already noted, Dirac was the first man in the history of science to make this profound discovery. For this he must be congratulated, and so must the modern scientific methods which have brought us so much practical information about ourselves and the universe in which we live.

Many centuries before Dirac, however, a text was revealed to an unlettered man in the Arabian desert. Its primary message was of the existence of one Creator and of an Afterlife in which human beings would account for their conduct while on earth. But this was no ordinary religious book, for it continually exhorted its readers to examine the natural world around them for the signs of God, to develop logical and objective thinking, and to place a crucial emphasis on education. This Book broke down the artificial boundaries which divide material and spiritual study. In the Holy Qur'an, science and religion are entirely compatible. Indeed, they cannot be distinguished.

In the Qur'an, we find the following verses, which may be considered relevant to Dirac's discoveries of the pairings and symmetries of the universe's structure, and in fact much more:

> Glory to God! Who created all things in pairs; those that the earth produces, as well as their own (human) kind, and things of which they have no knowledge. (Ya Sin 36:36)

The line "Who created all things in pairs" clearly shows that the Qur'an is referring to a law of nature; all things on earth and in the heavens—animal, vegetable or mineral (and even sub-atomic)—have been created in pairs. (Modern science has evolved around the study of pairs, from the male/female pair which generates animal and plant life to the pairings of quarks and leptons which are the most basic building blocks of the universe.) The line "Things of which they have no knowledge" suggests that the theory of pairs is of universal scope. It applies equally at all points in time and space; even in worlds of which we have no knowledge.

Elsewhere in the Qur'an we find a statement which modern science has spent its history confirming: "Verily all things (without exception), We have created, In exact measures and proportions". (Al-Hadid 57:49)

The line "In exact measures and proportions" is most revealing from a scientific point of view. The Qur'an contains the unambiguous lesson that as well as having perfect symmetry, every force in the universe is definite, of exact measurement and proportion following precise laws: "and there is not a thing, But its storehouse is with Us, And We send them down, But in definite ascertainable measures." (Al-Hijr 15:21)

God, the Absolute Singularity

The Holy Qur'an tells us of God: "He is the First, And the Last, He is the Outermost, And the Innermost, And He is the knower of all things." (Al-Hadid 57:3)

In this verse "the First" and "the Last" relate to the time domain, and "the Outermost" and "the Innermost" to the space domain. As such time and space are the first pair of nature. By extension all pairs are finally unified in the only Singularity in the universe—God. All pairs are aspects of the same reality, to which they all must return—the oneness of God.

But the Qur'an does not only give us information about the universe. It goes a step further, requiring us to use this information to develop our thinking. As always, the Qur'an demands the active participation of its reader: "And of everything, We have created in pairs, That you may reflect." (Ad-Dhariyat 51:49)

The Qur'an asks us to think, offering both a mental challenge and a proof of God's existence. If we reflect on the paired structure of the universe by using the methods of scientific reductionism, breaking a complex system down into its simplest units, we will finally come back to a single point of origin, an absolute singularity, the one source of time and space: God (Mahmood 1987). As the Qur'an says, "and all that is in the heavens, And all that is in the earth, It is from God; And (in the final analysis) Everything will return to Him." (Al-Imran 3:109)

Guidance for the Modern Age

The Qur'an comments on many areas of interest to modern science, from medicine to geology. Many verses of the Qur'an have taken on added significance as new scientific discoveries have been made. In fact, it seems that science has been following the Qur'an's lead for centuries.

Furthermore, in an age when progress in such fields as genetic engineering and chemical weaponry has raised complex ethical questions, the Qur'an teaches the salutary lesson that science should be the servant of humankind, not vice versa. It provides a model for all our endeavors in which symmetry is a prime value. Technological progress must be balanced by a symmetrical spiritual development. The new freedoms that we have won through our expanding knowledge must be balanced by an awareness of our new responsibilities. Otherwise our achievements will turn upon us, polluting our environment and increasing rather than diminishing our suffering. Without the guidance of Islam through the Qur'an, humankind with its scientific achievements will be like a child surrounded by dangerous toys.

This Book provided the light by which a great scientific culture once thrived. The Muslim world today, cursed as it is by illiteracy, intolerance and superstition, is a poor reflection of that former glory. This is the fault of human beings, not of the message of Islam, whose light continues to shine. Its light is every bit as available to a non-Muslim as it is to someone who has grown up in a Muslim environment. Once we move beyond our habitual beliefs and prejudices, all of us can profit from the message of the Qur'an. An independent and freethinking mind is all that is necessary to see its light. If we read, discuss and reflect upon the Qur'an, we can use that light to guide us now in our explorations of the universe and ourselves.

For this reason, we appeal to you to read the Qur'an. The examples of the Qur'an's truth presented in this paper are only a starting point. The Qur'an reveals further riches with each reading. The light of this Book, which spoke to seventh-century Arabs of things of which they had no knowledge, can teach modern humankind too.

References

Abdus Salam. *Symmetry Concepts in Modern Science*, Lahore: Atomic Energy Centre, 1966.

Bashir-ud-Din, Mahmood. *Doomsday and Life After Death*, Islamabad: Holy Qur'an Research Foundation, 1991.

Bashir-ud-Din, Mahmood, "Symmetry of the Universe and the Qur'an," *Science and Technology in the Islamic World*, Vol. 5, No. 1, Jan. 1987. National Science Council of Pakistan.

Bucaille, Maurice. *The Bible, the Qur'an and Science*, New York: Tahrike Tarsile Qur'an, Inc., 2003.

The Holy Qur'an: with English translation and commentary by Yusuf Ali

Time and Beyond as a Dimension

■ Osman Çakmak

E ven if we cannot easily grasp the real nature of "time," we can understand its aspect of being a "dimension." For example, specifying only a place without specifying a "time" for an appointment would not be sufficient. Let us presume that we are on board a space vehicle or a helicopter and we are announcing our present location by giving the ground coordinates, that is, the latitude, the longitude and the height. We have to specify our current time, that is, the date and the hour, in order to make such an announcement meaningful and proper. Space–time is thus a four-dimensional measurement system, the dimensions inseparable from each other, like the nail and the quick of a finger.

We certainly fail if we try to consider time as only a matter of determining the hour. It is, in fact, a dimension like depth, height and length. One reason for our difficulty in perceiving time may be caused by the fact that our optical perception is sensitive only to three dimensions, but no others. Many animals cannot comprehend the dimension of depth. Some animals see their environs in two dimensions as in pictures. We have difficulty perceiving other dimensions just as animals which see the world in two colors live without any awareness of other colors.

Humankind, with the most sophisticated senses, has a very different and privileged position above all creation. In spite of this, we have limited sight, hearing, and other senses. Many a world that is beyond our senses remains imperceptible to us.

Another aspect of time that supports its dimensional characteristics is that it is in full conformity with and proportional to other dimensions. In terms of its extent, the duration of events increases or decreases in parallel with spatial dimensions. Humans live for around sixty to seventy years, while microscopic animals live around one or two days. The life of the sun and the universe which constitute the cosmos is expressed in billions of years. On the other hand, the life of subatomic particles is expressed in billionths of a second. Thus, we assume them to be reso-

nances. There is time reduction together with and compatible with space constriction on the subatomic scale, and this fact is yet another proof that time is also a dimension.

How shall we understand the other dimensions of space? What does the fourth dimension of space mean? It is not easy to even imagine this, let alone describe it.

If "a" is the length, a^2 is the area and a^3 is the volume of a thing, then what is a^4? If we see space as a giant plain sheet of paper, that sheet of paper has no depth but only a surface. If we crumple it into the shape of a rough sphere, we obtain "Riemann space." Just like we perceive the three-dimensional earth as a two-dimensional surface while we are on it, this 3-dimensional sphere made of 2-dimensional paper will be perceived as 2 dimensional by us. We can only talk about the third dimension after we generate a depth, that is, after we step outside the paper and move above and below it.

The fourth co-ordinate of space is a tunnel. Let us suppose that the universe is two-dimensional, that is, it is like a thin sheet of paper, and let us human beings be like pictures with no thickness over its surface just like the pictures on a newspaper. We are free to move in all directions on this sheet of paper. We can sense four directions, but we will never perceive the terms "up" and "down" (or "upper" and "lower") since we will never leave the surface of this sheet of paper. Such terms will seem ridiculous to us even we are told of them. Accordingly, we will never hear of a third dimension and our vocabulary will never contain such terms as "up" and "down."

If a three-dimensional object existed above our fictional paper realm and if this object slit our paper realm, we still would not see it in three dimensions but only the part of it intersecting our paper realm. If such a thing were a sphere, for instance, we would see its projection in a circular form. Its latitudinal sections would gradually expand starting from the poles, reach their largest on the equator line and its ring-like (circular)  shape would gradually decrease and finally disappear at the other pole. That is, we would see it only as its cross section, or two-dimensional shadow. Such a three-dimensional object would seem two-dimensional to us since we would suddenly see its cross section. The sudden appearance, expansion, decrease and final vanishing of that spherical object In our two-dimensional realm would seem quite amazing to us since our shapes are fixed and immovable.

The three-dimensional shade of an extraterrestrial four-dimensional object overshadows our three-dimensional space. We see the linear tunnels in cross-section, not longitudinally, just as we see the sphere as circular. Though the sphere is a simple object, it amazes us.

Let us now imagine a more complex form. Let us, for instance, reflect the shadow of a vase onto a wall and obtain various shades by turning it repeatedly. A fixed and immovable portrait on the wall would regard the shadow and its variations reflected over the same plane with surprise and fear, since that portrait, or that person without depth, sees only what is reflected on the wall, but not us and the vase. The wall is the only realm for him and there is nothing for him beyond and behind the wall even if we say so.

We humans tend to assess events within the narrow limits of space and within certain dimensions, since we are bound within a single space–time cone. The conceptualization of space with its dimensions of height, length and depth is possible for us. However, the fourth dimension, time, is an abstract and metaphysical measure even though it is studied within physics. The tunnel thus seems to us an incredible dimension.

Our perceptions with the five senses in the visible universe can be considered as the projections of non-physical and multi-dimensional realities (the eighteen thousand realms) to our domain. Clearly, in order to gain a better understanding of those realms, which we do not see but which we feel exists, with the help of physics, we need to emancipate ourselves from the narrow patterns of time and space in this world of trial. We need to travel toward the horizon of spirit and develop an all-new scientific language which approaches physics and metaphysics together. Finally, we can say, in Bediüzzaman's words, that the physical and observable universe which is the domain of research for modern physics is an ornamented curtain veiling the world of the unseen.

A Tale of Design and Love

The value of the iron (or any other material) from which a work of art is made differs from the value of the art itself. Sometimes they may have the same value, or the art's worth may be far more than its material, or vice versa. An antique may fetch a million dollars, while its material is not even worth a few cents. If taken to the antiques market, it may be sold for its true value because of its art and the brilliant artist's name. If taken to a blacksmith, it would be sold only for the value of its iron[13].

E ach creation is a work of art. All animals and plants, as well as every human being, are unique and priceless. And those who appreciate their value are like antique dealers as in the passage above. I recently had the chance to listen to such an "antique dealer," Joanna Aizenberg of Bell Laboratories/Lucent Technologies, and witnessing the appreciation of the valuables she presented to us helped me better understand Said Nursi. Both the valuable object she was talking about and her appreciation of it were equally inspiring for me, and this is the reason why I have decided to share this story with you. Without any further ado, here is the story of a sponge species called the Venus' Flower Basket and its "eternally" incarcerated residents: a pair of shrimp. Now, you must find what is hiding behind all this; after all, it is the eyes that look but the heart that perceives.

Venus' Flower Baskets (Figure 1a) are vase-like sponges that grow upright on the sea floor of the Pacific Ocean, mostly around Japan. They have a very sophisticated mesh structure which caused medieval Europeans to assume they were glasswork made in China. In Japan they are called Kairou-Douketsu (together for eternity) and given as wedding gifts, since they generally house a pair of mated shrimp which are trapped in their cavity. As you have probably already understood, our story is about the engineering secrets of these sponges and their relationship with their guests.

---

13   Nursi, Said. *The Words*, Twenty-Third Word, First Point.

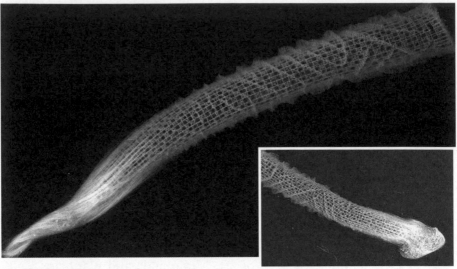

The Design

The skeleton of the Venus' Flower Basket is made of silica, which is a very brittle material (remember the glass windows that you broke with your football when you were a kid—they were made of silica). How can these amazing creatures withstand the pressure and the currents present at the sea floor or the disturbance caused by two shrimp? The secret lies in the hierarchical construction of their cylindrical cage-like structure. As can be seen in Figure 1b, their skeleton is made up of beams that run perpendicular and parallel to the axis of the sponge, which forms a rectangular grid. This grid is further supported by beams that run diagonally in both directions. Finally, this whole structure is reinforced by ridges that spiral around. But these are just the macroscopic hierarchical levels of the construction. Now let's start from the very first level of this hierarchy and try to understand how each level adds to the stability of the sponge.

The basic building block of the Venus' Flower Baskets is a fiber composed of silica nano-spheres (Figures 1i and 2a) that grows around an organic filament (the black dots at the center of the circles in Figure 1f). Though this fiber is not very stress tolerant, due to the size of the spheres from which it is made, in the next level of hierarchy it is toughened by alternating organic and silica sheets that form a concentric lamellar (fine, alternating layers of different materials) fiber structure. The thickness of each layer in the fiber decreases from 1.5 (: 1/1000 mm) at the center to 0.2 towards the periphery (Figures 1f, 1g and 2b). Hence any crack that is initiated at the periphery is halted at the organic interlayers and while the thinner outer layers lessen the depth of crack propagation; the thicker inner layers enhance mechanical rigidity. And in addition to their mechanical stability, these silica fibers are endowed with optical properties which are superior to man-made fibers.

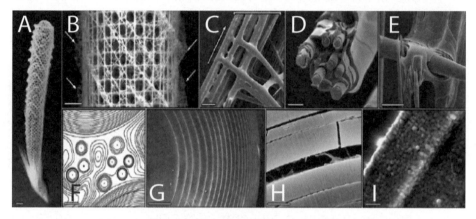

Figure 1. Structural analysis of the mineralized skeletal system of Euplectella sp.

(a) Photograph of the entire skeleton, showing cylindrical glass cage. Scale bar, 1 cm.

(b) Fragment of the cage structure showing the square-grid lattice of vertical and horizontal struts with diagonal elements arranged in a chessboard manner. Orthogonal ridges on the cylinder surface are indicated by arrows. Scale bar, 5 mm.

(c) Scanning electron micrograph (SEM) showing that each strut (enclosed by a bracket) is composed of bundled multiple spicules (the arrow indicates the long axis of the skeletal lattice). Scale bar, 100 mm.

(d) SEM of a fractured and partially HF-etched (25) single beam revealing its ceramic fiber-composite structure. Scale bar, 20 mm.

(e) SEM of the HF-etched (25) junction area showing that the lattice is cemented with laminated silica layers. Scale bar, 25 mm.

(f) Contrast-enhanced SEM image of a cross section through one of the spicular struts, revealing that they are composed of a wide range of different-sized spicules surrounded by a laminated silica matrix. Scale bar, 10 mm.

(g) SEM of a cross section through a typical spicule in a strut, showing its characteristic laminated architecture. Scale bar, 5 mm.

(h) SEM of a fractured spicule, revealing an organic interlayer. Scale bar, 1 mm.

(i) Bleaching of biosilica surface revealing its consolidated nanoparticulate nature (25). Scale bar, 500 nm. (Figure and captions from ref. 2.)

Fibers of different diameters reinforced this way are then bundled loosely in a silica matrix (Figure 1d and 1f). The different diameter of the fibers in the bundle and the weak lateral bonding between them are essential for increasing the strength of the bundle against crack propagation. At the next level of hierarchy, these bundles are used as building blocks of the cylindrical cage of the sponge, being arranged horizontally and vertically into a square grid. This grid in turn is reinforced by diagonal bundles that run in both directions along every second square lattice. The minimum number of pin-jointed struts (i.e. ones that are free to rotate at the joints) per node needed in order to form a rigid two-dimensional grid has been shown to be six; this is the number present in the skeleton of the Venus' Flower Basket. In fact, if the diagonal bundles were to run along every square lattice, the number of struts per node would be 8, which would be redundant for the stability in the skeleton.

At the early stages of the growth of the Venus' Flower Basket the struts are not connected at the nodes. However, as the sponge gets older the struts are joined by a silica cement which itself also has a lamellar structure (Figure 1e). Hence, while the younger sponges are flexible, the older ones are stiff; this also has important implications for the symbiotic relation that the sponge has with its guests, the shrimp. (This issue will be discussed in detail when the lifecycle of the shrimp is examined.) While the resulting grid is stable in two dimensions, in three dimensions it may still suffer from exterior effects, such as ovalization. This problem, however, is solved at the next level of hierarchy by the helical ridges that surround the grid (Figure 1b). The absence of the ridges at the base of the skeleton of the sponge where the cage diameter is small, and their increased density further up the cage where the diameter is much greater, is proposed as evidence supporting this argument. Finally, this whole cage structure must be anchored to the sea floor in a way that will withstand the bending stresses caused by the currents. This is managed through the use of the fibers that have been discussed earlier; they are used as connectors between the base of the sponge that is anchored to the sea floor and the vertical struts of the skeleton, resulting in a flexible connection that enables the cage to swing freely in the currents (Figure 1a).

As a conclusion, it can be said that "the resultant structure might be regarded as a textbook sample in mechanical engineering, because the seven hierarchical levels in the sponge skeleton represent major fundamental construction strategies, such as laminated structures, fiber-reinforced composites, bundled beams, and diagonally reinforced square-grid cells to name a few."

Now let's concentrate further on the fibers (or spicules) that anchor the cage to the sea floor. These anchorage spicules (a term used for describing the skeletal structures of sponges which comes from the Latin word speculum, meaning the head of a spear or arrow)[14] are 5-15 cm in length and 40-70 um in diameter. In the above discussion we have briefly discussed the cross-sectional structure of these fibers that gives them their flexible but resistant nature. Here we will focus on the optical prop erties of these spicules. But before doing so, let's briefly explain how optical fibers work.

Optical fibers are silica fibers of 5 to 80 um diameter that are coated with a cladding layer; light waves can travel in these for long distances by constantly bouncing off the cladding. The reason for this is the refractive index difference between the silica core and the cladding layer. Refractive index (n) is a measure of the ability of a medium to change the phase velocity of light and cause the light waves to bend while leaving one medium and entering another (refraction); in the case of fiber optics, leaving the core and entering the cladding. However, if the refractive index of the second medium is lower than that of the initial one, the incident light waves that have an incidence angle higher than a critical

---

14   Also defined as one of the minute calcareous or siliceous bodies that support the tissue of various invertebrates (Merriam-Webster's English dictionary)

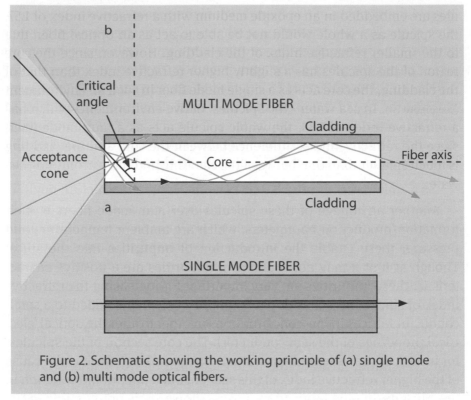

Figure 2. Schematic showing the working principle of (a) single mode and (b) multi mode optical fibers.

value or critical angle can be reflected back to the first medium and this is what happens in fiber optics (See red ray in figure 2). If the core diameter is small (5-10 um), light rays can propagate only through a single path in the fiber (which runs parallel to the fiber axis), hence these type of fibers are called single-mode fibers (See Figure 2a). If the core diameter is larger, however, (60-80 um) several paths are accessible, and more paths will have incidence angles that are greater than the critical angle; hence they are called multi-mode (See Figure 2b).

Now with this information in mind, let's have a look at the characteristics of the anchoring spicules of the Venus' Flower Basket. First of all, as mentioned in the previous discussion, the lamellar structure of these spicules prevents crack propagation, which is the main failure mode of commercial silica fibers. This lamellar structure, however, also determines the dependence of the optical behavior of the spicules on the environment in which they are embedded. For instance if the spicules are embedded in an epoxide medium with a refractive index of 1.57, the spicule as a whole would not be able to act as an optical fiber, due to the smaller refractive index of the cladding. However, since the core region of the spicules has a slightly higher refractive index than that of the cladding, the core acts as a single mode fiber in such an environment (see Figure 2a). In sea water—the spicules' native environment—which has a refractive index of 1.33, the whole spicule acts as a multimode fiber, since the refractive index difference between the core and the cladding is much smaller than that between the cladding and the surrounding sea water.

Another advantage of these spicules over man-made fibers is their formation/production parameters, which are ambient temperature and pressure; these enable the introduction of impurities into the silica. Though at first it may not sound as if impurities are a positive characteristic, these impurities are very important for increasing the refractive index of silica, and act as dopants (impurity elements added to a semiconductor lattices in low concentrations in order to alter the optical/electrical properties of the semiconductor). The core section of the spicules, for instance, shows increased sodium concentration, which is the cause of the higher refractive index of this section. Such dopant introduction in the silica during the fabrication process, however, is not possible in the case of man-made fibers, due to the very high processing temperatures.

In addition to this, the spicules have crown-like caps at their base and thorn-like structures throughout their middle section. While the crown-like termini most probably are used to anchor the sponge to the ocean floor, it has also been shown that the wave-guiding efficiency of the spicules increases when the illumination comes through the end that has the crown-like structure. Hence, it has been proposed that this structure may be acting as a light harvesting lens. The thorn-like structures, on the other hand, share the lamellar construction of the spicule body, and the light guided through the body branches out to these spines and emerges at the tip. Since sea water comes into contact with the tip at an almost perpendicular angle to the guided light, the coupling is efficient. Hence the combination of crown-like ends and thorn-like structures forms optical networks that collect and distribute light. However, at the depths inhabited by the Venus' Flower Baskets there is no accessible light source.

If one accepts the fact that there is no waste in nature—whether one believes in "creation" or "evolution"—the existence of such an advanced network-like structure as a part of a sponge—the most primitive animal—is at least thought-provoking. In the case of sponges that dwell in shallower waters with similar spicules, it has been postulated that such spicules gather and provide sunlight for the sponge's endosymbiotic algae. However, at the depths at which the Venus' Flower Baskets live, direct sunlight is not available.

However, it has been suggested that if light sources such as bioluminescent microorganisms (bioluminescence is the production and emission of light by a living organism as the result of a chemical reaction during which chemical energy is converted to light energy) or chemiluminescence (emission of light as the result of a chemical reaction) exist, their light may be efficiently distributed by the sponge and act as an attractant for juvenile shrimp that are searching for a host. But for now these suggestions are just speculation and merit further investigation.

Before concluding this section, we should also note that, as a natural outcome of their construction/composition, these spicules do not have as great a transparency as their industrial counterparts and light cannot be transferred over long distances with them. However, it seems this is not a problem for the Venus' Flower Basket as, apparently, they just need fibers of 5-15 cm to survive and it is the scientists who need to figure out a way to incorporate the traits of the Venus' Flower Basket into industrial fibers.

The Love

As mentioned in the introduction, the Venus Flower Basket hosts a pair of mated shrimp. These belong to the family of Spongicolidae, the Spongicala japonica. These shrimp, which can be as "big" as 9 mm in length, spend most of their lives in their host sponge. Though studies about them are limited, it is believed that before permanently being entrapped in their host, the shrimp have two free-living periods. The first one is just after hatching when they are small enough to exit through the mesh of the sponge. During this period they exit and re-enter their cages and live in a group with their parents and other juveniles. Studies suggest that the females generally stay with their parents until sexual maturity, whereas the males tend to leave their original host and live a solitary life until they reach a length of about 4 mm.

The second free-living period comes at the time of sexual maturity, when it is believed that the male and female mate outside and then invade a host, or the female searches for a host that is already occupied by a solitary male. During this stage, the shrimp have a body length of 3.5 to 6.5 mm, which is bigger than the mesh size of the host sponges. Though this seems puzzling, it is thought that the mated shrimp enter the sponge in its flexible stage—when it may be easier to penetrate through the mesh—and get trapped there "forever" as the sponge grows older and stiffer. In fact this theory is supported by the finding that several flexible sponge specimens host solitary and young mated shrimp, whereas in the stiff specimens only very few solitary and young mated shrimp have been observed.

References

1. "Biological glass fibers: Correlation between optical and structural properties." J. Aizenberg, V. C. Sundar, A. D. Yablon, J. C. Weaver, and G. Chen, *Proc. Nat. Ac. Sci.* 101 3358 (2004).

2. "Skeleton of Euplectella sp.: Structural hierarchy from the nanoscale to the macroscale." J. Aizenberg, J. C. Weaver, M. S. Thanawala, V. C. Sundar, D. E. Morse, P. Fratzl, *Science*, 309 275 (2005).

3. "Fibre-optical features of a glass sponge - Some superior technological secrets have come to light from a deep-sea organism." V. C. Sundar, A. D. Yablon, J. L. Grazul, M. Ilan, J. Aizenberg, *Nature* 424 899 (2003).

4. "Skeletal growth of the deep-sea hexactinellid sponge Euplectella oweni, and host election by the symbiotic shrimp Spongicola japonica" (Crustacea: Decapoda: Spongicolidae). T. Saito, I. Uchida and M. Takeda J. *Zool.*, Lond. 258 521 (2002)

5. "Pair formation in Spongicola japonica (Crustacea: Stenopodidea: Spongicolidae), a shrimp associated with deep-sea hexactinellid sponges." T. Saito, I. Uchida and M. Takeda J. Mar. *Biol. Ass.* U.K. 81 789 (2001).

Physics of the Unseen

■ H. Hüseyin Erdem

"Ever since the beginnings of modern science, four or five hundred years ago, scientific thought seems to have moved humankind and consciousness further from the centre of things. More and more of the universe has become explicable in mechanical, objective terms and even human beings are becoming understood by biologists and behavioral scientists. Now we find that physics, previously considered the most objective of the sciences, is reinventing the need for the human soul and putting it right at the centre of our understanding of the universe!"

The last century has witnessed a new scientific approach with the development of quantum theory. The theory has been tested to such a degree that it has become the most tested theory out of all time. This is probably due to the fact that it is the most mind-provoking theory to date. Nevertheless, the new theory has passed all these tests and has been confirmed as being more complete in explaining the cosmos than any previous theory. Quantum theory has shown that the old approach of a mechanical universe was an oversimplification. One of the most important consequences of this is that quantum theory refutes the main foundations of positivist philosophy. This philosophy sees the universe consisting of what we can observe or measure, with everything beyond not being real. This denial also applied to knowledge that came from religions, and this resulted in the present conflict between religion and science. However, today even modern science says that the universe cannot be limited to what we observe. The basic principles of quantum physics show the possibility that most life is beyond the scope of our observations, and we have no way of knowing about them via physical means.

Although positivist philosophy dates back to the 16th century, it was August Comte who defined it in a systematic way in the mid-19th century. The Harper-Collins dictionary defines Positivism as "the view that all true knowledge is scientific." Positivism includes the view of reductionism which claims that everything in the universe, including astrophysical systems, complex biological systems, social movements, cultural values, and belief systems can all be reduced to simple physical and chemical events. Probably one of the most unfortunate outcomes of this approach was the questioning of belief systems with the tools of the scientific method. In one of his articles Fethullah Gülen says: "The massive influence of positivism and materialism on science and on all people of recent centuries makes it necessary to discuss such arguments. As this now-prevalent "scientific" worldview reduces existence to what can be perceived directly, it blinds itself to the far more vast invisible dimensions of existence."

Such arguments against religion that spring from materialism have gone worldwide, and all religious faiths have been questioned. Even the faithful has been confused by these arguments, consciously or unconsciously. Although scientific knowledge should be only one source of knowledge, it was considered to be the only source. In Huston Smith's words, this was a "blank check" to science.

It should be clarified that the early founders of both classical and modern physics did not perceive science in a positivist way. Copernicus and Newton at the birth of classical physics, and Einstein, Dirac, and Planck at the birth of modern physics, all had religious convictions and envisioned science as a part of knowledge. Einstein was even accused of being a theologian in disguise by some scientific historians. It was positivist philosophy which took advantage of scientific developments and used them against religion, resulting in the apparent conflict today. However, new developments in science have proven that the basic assumptions of positivism are no longer valid. Thus positivism should be nothing but an outdated ideology.

From Quantum Physics to Metaphysics

Quantum mechanical behavior emerges when one observes phenomena at microscopic scales. One of its novelties can be seen in that it offers a more comprehensive atomic model. The new atomic model has very important applications to our life, ranging from making lasers to producing computer chips. The early understanding of an atom was that

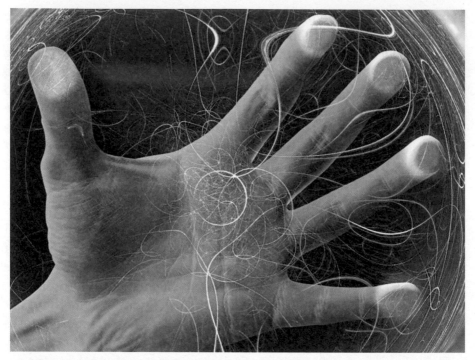

there was a nucleus at the center and electrons circulating around it, like in the planetary systems (the Bohr model). Although this was a great achievement at the time it was proposed, later scientists realized that classical physics cannot explain the circulation of the electron around the nucleus. In such a model the electron should lose energy and eventually collapse into the nucleus.

In the quantum mechanical definition the electron is more like a wave around the nucleus than a particle. So the electron is not really a particle orbiting around the nucleus, but rather more like a cloud that is spread evenly around. Sometimes the electron is called a particle because it acts like a particle in some experiments. As seen in this example, in a quantum mechanical measurement we cannot find an answer to "what the electron really is," but rather we find an answer to "how it responds to a particular setup." The actual stuff is a neither a particle nor a wave. We are rather measuring a form of its behavior compatible with our experimental system. Then, according to the quantum theory, there is no a way to completely understand this actual stuff with measurements.

Above we gave the famous measurement problem, which forms the heart of the quantum theory. Although what we are dealing with looks like a physical problem, "the measurement problem" has far-reaching philosophical consequences. The basic problem is that we need to know what this actual stuff looks like so that we can have an answer to the question of "what it really is." However, any explanation should be able to explain the transition from a quantum physical system into the macroscopic system in which we live so that we can have a meaningful model. Otherwise, paradoxes are inevitable (for further analysis see the famous Schrödinger's cat thought experiment).

The most complete and satisfying answer comes from the Copenhagen interpretation. It was proposed by Neil Bohr, one of the prominent figures in the development of the theory. Debates lasting for months or longer, especially between N. Bohr and A. Einstein, ended up with the victory of Bohr's ideas. According to the Copenhagen interpretation, the actual stuff is neither a wave nor a particle but is something not physical; rather it exists only in knowledge. This knowledge collapses into a physical state when somebody measures it. So the new theory suggests a very abstract approach to the universe as opposed to the old mechanical model. The famous astrophysicist Sir James Jeans wrote that "the stream of knowledge is heading towards a non-mechanical reality; the Universe begins to look more like a great thought than like a great machine. The human mind no longer appears to be an accidental intruder into the realm of matter."

Scientists think that the true picture of the actual stuff can never be completely understood in this physical universe because we are limited by our physical tools. There may be other states, but we have no tools to understand them or get to know them since we are limited by the tools of this universe. This is the point where the new physics talks about other dimensions which are beyond the observable and measurable universe. But this is exactly what philosophy calls "metaphysics." So we see that the new physics not only accepts the existence of other metaphysical realms, but it even says that they must exist for completeness!

The Necessity of Human Consciousness

A concern comes to mind about what is unique in this measurement process that results in the ultimate transition from a knowledge system into a physical system. How can the detector in an experiment result in this transition? The answer from the Copenhagen interpretation is very surprising. The detector cannot be the cause for this transition, because it does not make any changes in the system before or after the measurements are carried out. That is, these tools we use to make the measurements do not change anything in the nature of the system. Not even the eyes of the observers or the brain that is making this measurement can do this, as they are no different than the experimental apparatus, except that they are more complex. They are just part of the experimental system in this chain, like mechanical detectors. The chain continues until it ends up in human consciousness, which is something non-material, as any physical identification would put it in the same category as the previous members of the chain. Then the unique role of human action enters the system; measurement is part of the knowledge in the mind. With this measurement, human consciousness becomes aware of it. This is the unique property that the human being has, which cannot be attributed to any other objects and which plays a central role in the interpretation of the quantum theory.

We infer that human consciousness is something immaterial and behaves quite differently than any other entity in the universe. Interestingly, the distinction of the physical and spiritual side of human beings is found in the teaching of religions, which we now see in the context of modern physics. This is a very important reconciliation between science and religion and it is also reassuring that we are not like any other objects in the universe!

Is Materialism Coming To the End?

With the new developments in physics, a materialistic worldview seems to be a simplistic look at life and existence. We remember the classic statement of materialistic philosophy, "I only believe what I can see or measure in the laboratory." Quantum physics would respond to this by saying, "it is not that simple!" We see that there are no contradictions between the new physics and the teachings of religions. We do not know how God creates life in hereafter, hell, and heaven. But one thing we do know is that their existence does not contradict the modern scientific worldview. The realms of invisible creatures (like angels and the devil) and their interactions with our physical world cannot be understood with science. Modern physics says we should not seek knowledge of these through science. They can only be known by what is told to us in our holy books and by the prophets.

The extreme approach of materialism to the human being is that the human is the most evolved and most complex biological mechanism in the universe and in theory its consciousness and other feelings can be reduced to chemical reactions. This approach is in complete contradiction with modern physics. Modern physics says the human being is totally distinguished from other beings by our non-material consciousness. So we see that modern science removes the human being from the ignorance of materialism and puts us into the center of the universe. This is what religions have been saying since the creation of Adam and Eve!

As people of the twenty-first century, we can see that the discoveries of modern science do not contradict faith. We see that modern science is widening its horizons by identifying metaphysics as being part

of reality. In the words of the twentieth- century scholar Said Nursi, "... the radiance of conscience is religious sciences. The light of the mind is modern sciences. Reconciliation of both manifests the truth. The student's skills develop further with these two (sciences). When they are separated, from the former superstition and from the latter corruption and skepticism is born." A similar statement by Einstein is "...I cannot conceive of a genuine scientist without that profound faith. The situation may be expressed by an image: science without religion is lame, religion without science is blind."

A Dead End for Science or a Call to the Creator

■ Vitaliy Sheremet

T he scientists of the world have been engaged in solving the problem of deciphering the so-called human genome during the last
few decades. At the turn of the millennium the genetic map had
finally been deciphered; nevertheless, it could be said that classic genetics and all of the more recent research efforts in biology, biochemis-
try, physiology and some inter-disciplinary methodologies have found
themselves at a dead end. But this is only a dead end if we fail to recognize that we are all governed by a single supreme intellect, by the Divine
Providence, Who voices His Will by means of the Word.

The name of God is different in different languages of the world and
in the minds of those who accept the Creator as the only God: the Christians, Muslims, Jews and other believers, around sixty to seventy percent
of the global population. Yet this name corresponds for them with the Old
Testament, the New Testament, and the Qur'an. In all these instances,
there is a Holy Word in one form or another. The task of every believer is
to recognize the very form that the Creator uses to call His Creation. This
is a personal issue that originates from the religion that one practices.

So, what is the problem that faces genetics at the beginning of the
3rd millennium and how does it correspond with the Creator's expressed
will and His call to us?

It is here that there is a prospect for a remarkable breakthrough, if
only... This "if only" can be connected to academic achievements and
issues.

In fact, over the last three or four years scientists have discovered by
very sophisticated means and through careful research that the genetic
code that governs the human body—and, in a broader sense, everything
that is alive in Nature—accounts for no more than one percent of the
length of DNA molecules that determine the development of all living
species. This discovery was as shocking for scientists as the deciphering
of the genetic code had been. They concluded that genetic programming

occurred in the DNA molecular "free zone." Here, scientists—among them the Russian naturalists A.G. Gurevitch and V.I. Vernadskiy, who some fifty to seventy years ago claimed that a purely materialistic understanding of the gene was the limit to which non-believing science could go—were proven to be right.

The new discoveries are most certainly related to the emergence of such sophisticated physical instruments as the laser, holography, sol tonics and even powerful computers. Modern technology has proved without a doubt that the program in space and time for the creation of the human organism is not based on random accident, but rather is predetermined from "above." The protein molecules and the amino acids that comprise the gene (to date, more than twenty different types of amino acids have been discovered) are placed in a particular order. A single fitting lock-and-key relation exists in the composition of the genetic code components. In addition, it has been proven beyond a reasonable doubt that the genetic code of species that live on the Earth has not changed in three billion years; i.e., there is no room to talk about evolution, the principal postulate of materialists. Then, who or what has created everything that exists today on our planet several billion years after the creation of the Earth?

Then there is another puzzle: Why does the genetic code have such a small place, only taking up one percent of DNA?

Scientists in Russia have learned that ninety-nine percent of DNA, which was previously considered to be useless, hides within itself the so-called "genetic computer" that comprises the programs needed to make living organisms into a variety of species; and these mask the genetic features that are unique to a particular species. It is not completely clear how the mechanism of this so-called genetic computer works, but it does work. The concept of a holographic mechanism for the storage, transference, and recovery of information was developed as the result of an experiment.

Scientists took a freshly cut leaf, chopped off part of it, and put this between two slides and two photo plates. As the picture developed, it became clear that the leaf was depicted whole. In short, an idea or a phantom had been photographed. These first experiments were conducted in Russia in the 1960s.

Russian, American, and British genetic scientists continuously repeated the experiment, taking phantom photographs of different objects, and came to the conclusion that science was dealing with a multi-dimensional picture of the leaf, or its hologram.

Based on this, some other puzzles were solved. The "genetic computer" manages the development of holograms by means of special static waves, called sol tones (sol tonics is a special scientific branch derived from this term) that function in the DNA embryo cells.

Scientists have long since established that out of one single fertilized ovule other ovules start to instantly develop, as if on command; these are responsible, for instance, for making bone, muscle, nerve and other systems within the human body. And over this material process there floats an utterly immaterial phantom that dictates and shows the embryo the way to develop.

In other words, there is a certain image according to which development proceeds. The DNA is the text that controls this creation, with its inherent rules of composition; it is possible to see DNA as being made up of letters, i.e. a word. At first there was the Word! This is a quote from the Bible. In Islam, Almighty Allah gave the Word by means of

the Qur'an (reading) to Muhammad. A phrase from the Qur'an describes the above process in an amazingly simple and pertinent way: "It is He Who fashions you in the wombs as He will. There is no deity but He, the All-Glorious (with irresistible might), the All-Wise." (Al Imran, 3:6)

The Word of the Creator, according to which the genome "works," is registered with great security in the bio-system apparatus. It will only disappear in conjunction with the last of the human beings. This may be the very idea behind what is called the Day of Judgment, or Doomsday, in Islam and other religions.

In the context of Einstein's principles of a single field theory, as well as in Shipov's physical vacuum theory, it may be possible to find clarification of the phenomenon of a wave matrix (copy) remaining "clean" in one case, but distorted in another.

It is worth discussing here those things that have already been proven. The programs written in the DNA cannot have emerged as a result of simple evolution, simply because of the huge volume of information contained here. The time required would have exceeded the time that the Universe has existed, around fifteen billion years. We have established an approximate time that would be required for the genetic transformations that determine the essence of human beings to occur. It is substantially less...

Another study has been carried out that does not fit into the traditional materialistic frame. It seems that the internal structural information of DNA alone is not enough to develop an accurate replica of the image organism from the composition of the protein elements. Numerous experiments carried out by Russian scientists (in particular, from the Moscow Scientific and Cardiology Center) have proved that a frog embryo that has been purposefully protected to a great degree from external influences will begin to distort, suffer malformation and finally die. This means that DNA has to be connected—maybe by means of sol tones waves or other contacts still unknown to us—with an "external source" that guides the genome-bio-computer work from somewhere in Space. One cannot but recall Muhammad here, the last of the Greatest Prophets, who categorically rejected the possibility of not only seeing, but even imagining the Almighty.

There are few people who still argue about the existence of the soul. The time when the soul departs from the body has been well-documented by scientists, doctors, and naturalists. As a matter of fact, the soul emerges when the heart cells die, that is, when the organism as a whole dies. It is at this time that a certain phantom of the genetic apparatus is formed, similar to the one described above in the phenomenon of the phantom leaf. It is interesting to contemplate the idea that the phantom of a human genetic apparatus that has lost its life by force would possess a high biological reaction and would therefore be in a position to distort and destroy any healthy molecules that may be close by. One cannot but recall the command to not kill the innocent contained both in the Old Testament and the Qur'an, a call to leave retaliation to Him and to Him only.

In conclusion to my brief essay on the necessity of belief in today's science, the common scientific method tries to explain "how," but it fails to answer the question of "why" that lies behind the mystery of existence. Any scientific approach rejecting faith is doomed to fail, for faith is an inherent need for us. The belief in Him, the Single and Almighty, is genetically programmed. In a *hadith* reported in *Sahih al-Bukhari*, God's Messenger states that every person is born in the primordial nature (*fitra*) of Islam.

Here, at the beginning of the third millennium, at the height of our scientific achievements, we have come to understand God as a natural phenomenon. We must follow His guidance and not distort the Word or the Image of love that has been implanted by the Creator in our genetic code by mindless acts and evil speeches.

Ensoulment: When Does Human Life Begin?

■ Olgun Hasgül

O ne of the most controversial topics in modern bioethics, science, and philosophy is to try to pinpoint the beginning of an individual human life. The consequences of this discussion are vitally important, as they may help to articulate more adequate arguments on some bioethical issues, like the definition of the moral status of the embryo, abortion, and embryo research[15]. Many philosophers and scientists have argued about the definition of personhood and when the beginning of a human individual's life occurs, yet an acceptable explanation has not yet been provided. In this field there is a temptation to ask science to choose between opinions and beliefs, yet these neutralize one another. We believe that the question of when human life begins requires the essential aid of different forms of knowledge. Here we become involved in the juncture between science and religion, which needs to be carefully explored[16]. In this article we will therefore try to explore this issue from different perspectives. To begin with, we can agree, as pointed out by Mason and McCall Smith, that what constitutes the state of being a person, or personhood, is a moral decision[17]. Most religious traditions hold that what makes one a person is the possession of a soul[18]. When the body meets the soul, what results is a human person, with all the attendant rights, especially the basic right to life. In light of these considerations our duty is to search for the exact time of ensoulment, in order to prevent us from terminating the lives of actual "human persons," namely by abortion or embryo research. The purpose of the present study is therefore to determine when, in the course of normal development, a human being begins to exist.

The availability of embryonic stem (ES) cells isolated from human blastocysts may open novel avenues for medical treatment of otherwise

15 Aksoy, S., "Personhood: A matter of moral decisions." *Eubios J Asian Int Bioethics*, 1997, 7:3-4.
16 Kurjak, A., "The beginning of human life and its modern scientific assessment." *Clin Perinatol*, 2003, 30:27-44.
17 Mason, J.K., McCall Smith, R.A., *Law and Medical Ethics*, London: Butterworths; 1984.
18 Glannon, W., "Tracing the soul: medical decisions at the margins of life." *Christ Bioeth*, 2000, 6:49-69.

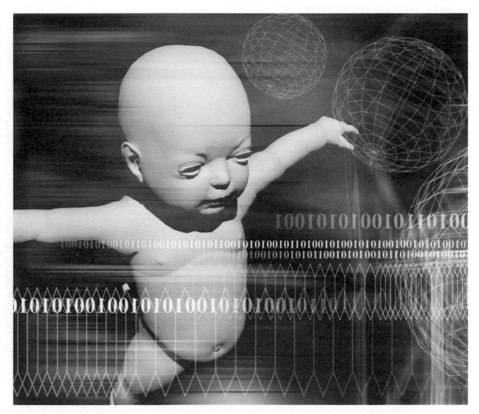

incurable diseases[19]. ES cells are pluripotent, apparently able to make any cell, except placental cells, and are also immortal[20]. However, the generation of human ES cells requires the destruction of early human embryos. This raises the same ethical questions and conflicts that are often heard when the ethics of abortion are discussed: Most people in the pro-life movement regard an embryo to be a full human person with a soul. It has all of the rights of any citizen, including the right to life. Thus any procedure that injures or kills an embryo is seen as murder.

However, most people in the pro-choice movement regard the beginning of human personhood to occur much later in pregnancy. Thus, killing a recently fertilized embryo is not seen as the murder of a human

19 Waite, L., Nindl, G., "Human embryonic stem cell research: an ethical controversy in the US & Germany." *Biomed Sci Instrum*, 2003, 9:567-572.
20 Smith, A., *Embryonic Stem Cells in Stem Cell Biology*, Marshak DR. et al. (Eds.) New York: Cold Spring Harbor Laboratory Press; 2001: 2-19.

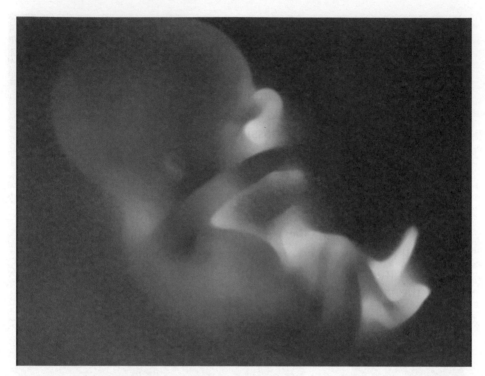

person[21]. What is the status of the embryo when it is several days old? Is the embryo alive? Yes, from its beginning the embryo is cellular and it is alive; no one questions this. But, is the embryo human? If we could catch the embryo before it reaches a stage where it is judged as human, we would be able to take the ES cells without any concern. In an earlier paper[22] it was argued that human life begins when the newly formed body systems begin to function as a whole, towards the end of the embryonic stage. Seeing that twinning can occur as late as day fourteen after conception and that such identical twinning will produce two individuals with different lives, this could be proposed as a pre-embryonic stage, i.e. the stage where the single individual person has not yet been determined. In this respect, The Ethics Committee of the American Fertility Society defines an embryo as distinct from a pre-embryo, based on medical science and legal precedents. According to that report and the Warnock Report, which lay behind the 1990 Human Fertilization and Embry-

21 Heinemann, T., Honnefelder, L., "Principles of ethical decision making regarding embryonic stem cell research in Germany." *Bioethics*, 2002, 16:530-543.
22 Shea, M.C., "Embryonic life and human life." *J Med Ethics*, 1985,11:205-209.

ology Act, the pre-embryo stage is considered to last until fourteen days after fertilization; four and the pre-embryo is to be respected, but not accorded absolute protection. Today, on the basis of these reports and subsequent legislation, embryo research is allowed up to fourteen days of development, up until the formation of the so-called primitive streak, or the beginning of the nervous system, after which the splitting and the forming of twins is no longer possible.

The Christian tradition on this subject is interesting, but can be read in two ways. In that tradition, abortion at any stage has always been re-garded as gravely sinful. However, for many centuries the termination of a pregnancy at an early stage carried lesser penalties than a later one. This was related to the view that the human soul did not enter the em-bryo until forty days or so after conception, an understanding that was taken over from Aristotle[23]. Thus, most Christians make a distinction be-tween the moral status of the unformed and the formed embryo, and think of the human person in the full sense coming only with a delayed ensoulment. For other Christians, however, fertilization is the point at which human life begins.

Human Life (Personhood) May Begin at around the 9th Week of Development

While we were searching for the beginning of a human individual life, we encountered a statement from God about the time of ensoulment (the infusion of a human soul in biological matter) in His last book[24]. God says in His book (the Qur'an) that, We made out of the "embryo" bones, and clothed the bones in "muscles'" (23:14). This statement in-dicates that in the embryonic stage, first bones and then muscles form. This is in accordance with embryological development. First the bones form as cartilage models at the seventh week, and then the muscles de-velop around them from the somatic mesoderm at the eighth week of development. Then We developed out of it another creature. This next part of the statement (23:14) implies that the bones and muscles result in the formation of "another creature." This may refer to beginning of

---

23 Folscheid, D., "The status of the embryo from a Christian point of view." *Ethics Med*, 1994, 10:57-59.
24 Moore, K.L., "A scientist's interpretation of references to embryology in the Qur'an." *JIMA*, 1986, 18:15-16.

the personhood (ensoulment) after the end of the eighth week. At this stage, the embryo has distinctive human characteristics and possesses the primordia of all the internal and external organs and parts, and after the eighth week, the human embryo is called a fetus. After He gave it (the embryo) the most appropriate shape, He gave it a soul and then He gave you hearing and sight. This statement (32:9) supports the statement mentioned above (23:14) in that after the embryo takes shape, the soul is infused into the fetus. This statement also indicates that the special senses of hearing and seeing develop in this order, i.e. after the ensoulment; this too is consistent with embryological knowledge. The formation of the internal ears occurs before the beginning of the eyes, at about the tenth and eleventh weeks of development, respectively. Moreover, it is very meaningful that God calls the fetus "you" after the beginning of personhood (ensoulment) (32:9), while the embryo before ensoulment is called "it" (23:14; 32:9) indicating the status of the embryo as only a "thing" or a cell cluster.

We suggest therefore that the beginning of a human person as an individual living organism is when the embryo develops into fetus at around the ninth week of development (after the fifty-seventh day) after the bones and muscles form, but before the development of hearing and sight. Only at this point do we have a multicellular organism and not merely a mass of living cells stuck together. The soul requires that there is an individuated matter present; prior to this period, there did not previously exist an individual human organism. A sperm or ovum that exists independently of the other only has a potentiality for human personhood. An embryo which is merely biological material that contains human DNA also has only a potentiality for being human personhood, which implies that it is not yet an actual person[25]. Two things may both have only a potentiality to be something else, but one of the two may be closer to actualizing that potentiality than the other. Thus, it could be contended that the embryo is closer to being an actual human person than the sperm or ovum is. Therefore, we can say that the informational capacity of the zygote and the early embryo is not sufficient to direct the development of anything personal, and is not sufficient to constitute a genetically stable subject as a human being.

25   Eberl, J.T., "The beginning of personhood: a Thomistic biological analysis." *Bioethics*, 2000,14:134-157.

Josef Seifert treats the human body and the soul as two incomplete substances (dualism) that are each completed in human beings as a compositum[26]. This position seems to require that the immortal soul is infused into material committed to being a human individual, and indeed into a fetus with sufficient cortical development to allow cognitive functions. This view is consistent with our suggestion that the soul is added to the already-existing physical body when the newly formed body parts and systems begin to function as a whole. If the soul takes effect when biological processes have produced a new human life, neither earlier nor later, then it would follow that ensoulment must occur at the end of the embryonic stage when, with the newly-formed brain acting as the central information-exchange point, the commencement of the functioning of the whole produces a new level of life and enables the processes that lead to personhood to begin. This alternative is compatible not only with the facts of modern medicine but also with the traditional understanding of ensoulment as defended by many Roman Catholic moral theologians, which has its roots in Aristotle. These moral theologians argue that the peculiarly human soul is not incarnated until there is an appropriately

---

26   Seifert, J., *Leib und Seele*, Salzburg: Universitatsverlag Anton Pustet; 1973.

organized matter[27]. However, some Christians believe that Jesus was a human being from the moment of conception and therefore that every human being must come into existence at the moment of conception. But some others believe that the conception of Jesus is an exception. It is believed that the events which occurred during the conception and development of Jesus will be explained by Jesus himself when he returns.

That the embryo cannot be considered an individual human being has implications for debates concerning the morality of morning-after pills, intrauterine devices which are abortifacient by stopping implantation, the disposal of excess embryos formed in the process of in vitro fertilization, and embryo research[28]. Both in the religious context and without it, abortion is difficult to justify at any stage after conception. But where the question of the possible use of surplus laboratory embryos is concerned, the time of ensoulment does matter, and if this does not occur until late in the embryonic stage of development, there seems no reason why the embryos may not be used, provided—as also seems to be the ethical conclusion even before the religious aspect is considered—it is for no purpose less than the relief of human suffering which cannot be

27   Bedate, C.A., Cefalo, R., C., "The zygote: To be or not to be a person." *J Med Philos*, 1989, 14:641-645.
28   Bole, T.J. "Metaphysical accounts of the zygote as a person and the veto power of facts." *J Med Philos*, 1989, 14:647-653.

relieved in any other way[29]. If "ensoulment" does not occur, as we have suggested, until the new organism functions as a whole, then a decision not to make use of laboratory embryos for medical purposes would be a grave responsibility.

The Soul May Have an Etheric Body

Science postulates that all matter is composed of atoms; atoms, however, are composed of protons, neutrons, and electrons, and those in turn are composed of still finer components, until we attain ether. This ether is a universal connecting medium, filling all space to the furthest limits, penetrating the interstices of the atoms without a break in continuity. So completely does it fill space that it is sometimes identified with space itself, and the universe is built up in this fluid and moves through a sea of ether[30]. The atom, the electrons, and the protons of which it is composed all move in a sea of ether, the very air we breathe, the very bodies we inhabit; all things likewise are moving in this sea of ether, the parent element from which all manifestation has been derived. By a kind of instinct, one feels it to be the home of spiritual existence, and to be the vehicle of both matter and spirit. No experimental data can be sufficient to bring us to the recognition of a soul, but there must be a substance that is the basis of personal identity, for without space-occupying substance, there would be no way to account for the soul's ability to move the body[31], and the idea of personality or a soul after bodily death would be inconceivable[32]. It seems possible that the substance of the soul could be ether. The modern concept of science is that ether is the primary form of all substance and that all other forms of matter are merely differentiations of ether; it then seems that the substance of the soul which in this life is linked organically with the body is identical with ether. The etheric (astral) body seems to be a supersensible element in humanity that primarily lives and acts in time, but also works into the dynamics of the life processes of the physical body. The soul is likely to work into the physical body of the human directly and exclusively via that etheric body.

29 Shea, M.,C., "Ensoulment and IVF embryos." *J Med Ethics*, 1987, 13:95-97.
30 Gonzalez de Posada, F., "Reflections on the ether." *An R Acad Nac Med* (Madr) 2001, 118:43-72.
31 Murphy, N., "Whatever happened to the soul? Theological perspectives on neuroscience and the self." *Ann N Y Acad Sci*, 2003, 1001:51-64.
32 MacDougall, D., "Hypothesis concerning soul substance, together with experimental evidence of the existence of such substance." *American Medicine* April, 1907.

The Limbic System Seems to Be Primarily Related to the Soul in the Brain

Humans have always contemplated the question of the anatomical location of the soul. The early concept that the cerebral ventricles harbor the soul began to break down only in the Renaissance[33]. During the Renaissance, the controversy crystallized into those who supported the heart (cardiocentric soul) and others who supported the brain (cephalocentric soul) as the abode for this elusive entity[34]. The brain seems to be the organ that harbors the soul, since the function of the brain is fundamentally linked to being human. The brain controls almost all the functions of the body and determines its psychological makeup, such as intellect and, in a theological sense, the soul. Without the brain, such functioning is not possible, since brain death means the end of human life. Children born with anencephaly can never experience a human life. Human identity, personality, and worth are associated with the functioning of the brain[35]. Considering the role of the brain in the maintenance of the dynamic equilibrium of the organism, there are compelling reasons for defining the brain as the organ which harbors the soul.

The limbic system is intimately concerned with emotional expression and with the genesis of emotions. The term "limbic system" is applied to the part of the brain that consists of a rim of cortical tissue around the hilus of the cerebral hemisphere and a group of associated deep structures—the amygdala, the hippocampus, and the septal nuclei. One characteristic of the limbic system is the paucity of the connections between it and the neocortex (the cortical tissue of the remaining nonlimbic portions of the hemisphere). It is stated that "the neocortex sits astride the limbic system like a rider on a horse without reins[36]." In fact, one of the characteristics of emotions is that they cannot be turned on and off at will. Since the limbic system is intimately concerned with the genesis of emotions and is critically involved in neuropsychiatric disorders[37], we can attribute to the limbic system the role of being a seat from which the soul can work in the physical body.

33   Schiller, F., "The cerebral ventricles. From soul to sink." *Arch Neurol*, 1997, 54:1158-1162.
34   Del Maestro, R., F., "Leonardo da Vinci: the search for the soul." *J Neurosurg*, 1998, 89:874-887.
35   Werning, C., "Is abortion murder?" *Med Monatsschr Pharm*, 1995, 8:251.
36   Ganong, W.,F., *Review of Medical Physiology*, Lebanon: Typopress; 1989.
37   Heimer, L., "A new anatomical framework for neuropsyciatric disorders and drug abuse." *Am J Psychiatry*, 2003, 160:1726-1739.

The Vomeronasal Organ Seems to Be the Point of Entry for the Soul and The Etheric Body

We believe that while the soul has an etheric component (substance), there must be an open window to the brain for the entrance of the soul with the etheric body. In this respect, the vomeronasal organ (VNO), which is found in the nasal cavity and which has connections with the brain only between the sixth and thirteenth weeks of human development—a period including the suggested time of ensoulment at the ninth week—seems to be the most appropriate window through which the soul and the etheric body can enter the brain. This suggestion is consistent with statements in the Old Testament (Torah) (Genesis-Bereishith 2:7, 7:22), which state that the life (soul) is infused into the human being through the nose. Actually, the VNO is said to be the place in the body where the nervous system is closest to the external world. Axons of VNO cells pass through the tiny foramina in the cribriform plates of the ethmoid bone to enter the brain[38], and to make close connections with the amygdala and limbic system, the seat of emotional, hormonal, and autonomic control; in short, the seat of the soul.

The vomeronasal organ is a fluid-filled, tubular structure located at the base of the nasal septum that opens into the nasal cavity via a duct at its anterior end. It is a chemoreceptive structure with direct axonal connections to the accessory olfactory bulbs in many terrestrial vertebrates[39]. Pheromones presumably bind to the vomeronasal organ and spur behavioral or physiologic responses, thereby allowing chemical communication between animals of the same species. The effects of pheromones are thought to be mediated by signals from the main and accessory olfactory bulbs to the amygdala and hypothalamus. The vomeronasal system, which is well developed and functional in adult animals, begins to function before or after birth in these animals[40]. The human VNO, however, seems to be different from that of animals in that it becomes rudimental before birth[41]. The VNO in the human embryo contains bipolar cells simi-

---

38 Moore, K., L., *Clinically Oriented Anatomy*, 2nd Ed. Baltimore: Williams and Wilkins; 1985.
39 Zbar, R.I., Zbar, L.I., Dudley, C., Trott, S.A., Rohrich, R.J., Moss, R.L., "A classification schema for the vomeronasal organ in humans." *Plast Reconstr Surg*, 2000, 105:1284-1288.
40 Salazar, I., Lombardero, M., Aleman, N., Sanchez Quinteiro, P., "Development of the vomeronasal receptor epithelium and the accessory olfactory bulb in sheep." *Microsc Res Tech*, 2003, 61:438-447.
41 Knecht, M., Witt, M., Abolmaali, N., Huttenbrink, K.B., Hummel, T., "The human vomeronasal organ." *Nervenarzt*, 2003, 74:858-862.

lar to the developing vomeronasal sensory neurons of other species, but the structure becomes more simplified later in development, having no obvious way to communicate with the brain. The first appearance of the tubular VNO is in the sixth week of human development[42]. This is when the human VNO resembles that of primates with functional VNOs[43]. An examination of the VNO and adjacent tissues suggested that the VNO may lose receptor cells and corresponding vomeronasal nerves and become a ciliated, pseudostratified epithelium at approximately the thirteenth week[44]. These observations indicate that

1) all embryonic humans develop a functional vomeronasal organ which is homologous with the VNOs of other mammals between the sixth and thirteenth weeks of age

2) the human vomeronasal organ does not degenerate prenatally, but very likely loses the functional components of the vomeronasal complex of other mammals; and

3) the remnant of the human VNO persists until birth and beyond[45].

These observations strongly support our hypothesis in that the VNO seems to have its main function only during the intrauterine period in humans, especially during the period of ensoulment. The VNO becomes vestigial before birth in humans while it begins to function before or after birth in animals. This difference seems to be very significant, since animals do not harbor any soul, unlike humans, and their VNO is important only after birth, when it allows them to communicate chemically with other animals of the same species. Although no anatomical connection has been demonstrated in humans, Monti-Bloch et al. deduce a physiological connection with the brain, because stimulus delivery to the VNO pit elicited several systemic responses[46], such as changes in blood pressure and heart rate, small but significant changes in hormonal

42 Bhatnagar, K.P., Smith, T.D., "The human vomeronasal organ. III. Postnatal development from infancy to the ninth decade." *J Anat*, 2001, 199:289-302.

43 Smith, T.D., Bhatnagar, K.P., "The human vomeronasal organ. Part II: prenatal development." *J Anat*, 2000, 197:421-436.

44 Witt, M., Georgiewa, B., Knecht, M., Hummel, T., "On the chemosensory nature of the vomeronasal epithelium in adult humans." *Histochem Cell Biol*, 2002, 117:493-509.

45 Johnson, E.W., Eller, P.M., Jafek, B.W., "Calbindin-like immunoreactivity in epithelial cells of the newborn and adult human vomeronasal organ." *Brain Res*, 1994, 638:329-333.

46 Monti-Bloch, L., Jennings-White, C., Berliner, D.L., "The human vomeronasal system." A review. *Ann N Y Acad Sci*, 1998, 855:373-389.

levels and some changes in mood[47]. Functional brain imaging studies also revealed consistent activation of the hypothalamus, amygdala, and cingulate gyrus-related structures during adult human VNO stimulation. These findings also support our view in that the VNO is strongly related with the emotional centers that harbor the soul, even in the absence of anatomical connections.

Conclusion

The above considerations make it seem likely that human life (person-hood) may begin at around the ninth week of development with a delayed ensoulment, and that the soul has an etheric component. The limbic system seems to be a primary center for the soul, while the soul may enter the brain with an etheric body, possibly through the window of the vomero-nasal organ, which is functional and has connections with the brain only during the suggested time of ensoulment. Therefore lawmakers, philosophers, scientists and any-one in a related field may consider the beginning of human life in their decisions and procedures as being from the ninth week (after the fifty-seventh day) of human development. Before this period, the embryo must be regarded as a cell cluster which is to be respected, but not accorded absolute protection. Our second conclusion is that embryonic stem cells have an enormous promise to benefit mankind—to save lives and cure or treat diseases which generate a very strong moral imperative to explore their potential. Almost all spare embryos in fertility clinics will eventually die, due to operator error or equipment malfunction. Spare embryos are routinely destroyed, by flushing them down a drain, by incinerating them, or by thawing them out and allowing them to die. They might as well have their stem cells extracted so that they can be of some use to humanity. If the above thesis is true, then there is no loss of human persons. What is destroyed in this process is human biological material that has not been infused with a soul. We can therefore explore the potential use of embryos and help people who are now burdened by debilitating diseases. So let us enter into the processes of nature and apply the knowledge so derived to help others, to alleviate pain, and to enhance human well-being.

---

47   Grosser, B., I, Monti-Bloch, L., Jennings-White, C., Berliner, D.L., "Behavioral and electrophysiological effects of androstadienone, a human pheromone." *Psychoneuroendocrinology*, 2000, 25:289-299

Two Different Fields for Science and Religion

■ Ali Ünal

C hristianity did not develop as a comprehensive religion encompassing all fields of life but as a set of spiritual and moral values with some bearing on, but no directions for, the "worldly" aspects of life. This has had serious consequences for subsequent Western history. For example, Christianity condemned war and, although war is a reality of human history, neglected to lay down rules for it. However, as this attitude never sufficed to end wars, the lack of "religious" rules and regulations about war has caused great brutalities in the wars which have taken place in the West and in the ruthless massacres by the Western powers throughout the world. Similarly, Christianity's condemnation of the "world" and nature as a veil separating man from God has been a major factor in encouraging the modern sciences to reject religious authority as irrelevant. Also, the sharp separation of this world and the Hereafter, religion and sciences, spirituality and physicality, led the thinkers and philosophers who tried to find a space for religion beside science to assign different fields to religion and science, reason and revelation, this world and the Hereafter.

Cartesian Dualism

The name of Descartes, the French mathematician and philosopher (1596–1650), is most famously associated with this dualism in Western culture. His ideas contributed to the almost complete separation of intellectual and scientific activities from religion and, in later centuries, to the Enlightenment, the mechanistic view of life, positivism and materialism. Cartesianism provided a shelter for those who searched for religion in life beside science; it also gave rise to many misconceptions about the relationship between life, religion and science. The intellectuals or philosophers who did not want to forsake either religion or scientific reasoning appealed to Cartesian dualism to justify their position. This manner of defending religion against scientific materialism still prevails among certain Muslim intellectuals. According to them, there is a world of qualities separate from a world of quantities. Science has the authority in the world of quantities and uses observation, measurement and experiment; while in the world of qualities, where observation, experiment and measurement do not apply, religion has the right to speak. So, religion can never be contradictory to science, but then religion and science have nothing to do with each other.

Cartesian Dualism Gives Science Superiority over Religion

Although intended to defend religion against science, Cartesian dualism gives science superiority over religion and primacy in practical life and thought, restricting religion to a set of blindly held beliefs not subject to research, verification and reasoning, and indeed practically irrelevant to the world and "worldly life." This attitude misrepresents religion as only a matter of believing or disbelieving, with the consequence that there is not much difference between accepting a "true" religion and believing in any religion whatever, even in myths and superstitions. It is this dualism which lies behind modern trends that see religion—without making any discrimination between God-revealed and man-made ones— as a set of dogmas inaccessible to reason, and quite cut off from science and the perceptible world.

However, religion, especially Islam as the last and perfected form of the God-revealed religions, demands, rather than believing blindly, both rational and spiritual conviction based on thinking, reasoning, searching and verification. Although it is acceptable to enter religion through the gate of imitation, it is never advisable to remain content with belief coming from imitating others. The verses in the Qur'an related to legal issues do not exceed 300, while there are more than 700 verses urging people to study "natural" phenomena, to think, reason, search, observe, take lessons, reflect and verify. Many verses conclude with "Will you not use your reason"; "will you not think"; "will you not reflect"; "will you not take lessons"; "take lessons, O men of insight". The Qur'anic condemnation of unbelievers as people having no intellects with which to think and reflect, no eyes with which to see and no ears with which to hear, are serious warnings for those who see religion as a set of blindly held beliefs and who are unable to discern the essential and unbreakable connection between religion and life, nature, reason and scientific activities.

Modern science takes the natural world as its field of study. The restriction of science by Cartesian dualism to the material, observable realm of existence, not allowing the possibility that there are other realms of existence and fields of study, may well be regarded as a way of keeping scientific inquiry "factual" and "objective." However, this attitude frequently leads to the view that the study of realms or subjects beyond the material, and the conclusions drawn from that study, are unscientific and therefore require neither research nor verification but only belief. It also carries many into agnosticism, to neither deny nor affirm the more profound and broader dimensions of existence beyond the material. A truly objective science should either accept that there may be many other truths and realms whose existence it is unable to discover by its present methods, or else change its tactics and techniques and equip itself with the methods necessary to discover those realms. As long as science persists with its rigidly empirical approach and methods, it will never be able to comprehend the full reality of existence. It is quite unfortunate for science that it reduces humans, as it does the universe, to mere physical existence, and tries to explain all intellectual and spiritual activities in wholly physical terms.

Modern science deals with nature as structured but aimless or meaningless concurrence of material things. Basically, there is not much difference between this attitude toward nature and Christianity's condemning it as a veil separating man from God. By contrast, Islam presents natural phenomena (but not the supra-natural ones whose existence and reality science either rejects or regards as unknowable by "scientific" methods) as evidence of its truth or reality and calls people to study and reflect on them and thereby harvest the nectar of belief. Nature, according to Islam, is the realm where God's Beautiful Names are manifested, and is therefore a set of "ladders of light" by which to reach God. Having originated from God's Attributes of Will and Power, nature is the "created" counterpart of the Qur'an, which originated in the Divine Attribute of Speech. So, nature is a book like the Qur'an, or it may be regarded as a city or palace, with the Qur'an being a sacred pamphlet explaining its meaning and how to dwell in and benefit from it. Human beings are the third counterpart of these two books, equipped with consciousness and will. This is why many Muslim scientists such as Ibn Sina (Avicenna), Zahrawi, Ibrahim Haqqi, Nasir al-Din al-Tusi and Ak Shamsaddin were practicing Sufis and well-versed in religious sciences.

Nature is an Exhibition of Evidences of Divine Unity

As nature has a sanctity because it is the result of the manifestations of the Divine Beautiful Names and a collection of mirrors reflecting the Divine Names and Attributes, the order and constancy in it are two significant proofs of Divine Unity. It is this order and constancy to which science owes its existence. The perfect order observed in the universe is the result of the fact that with all its parts and minutest particles, the whole of the universe is the work of a Single Creator. This is why there is an interrelatedness, co-operation and solidarity among all the parts of the universe and the creatures in it. For example, in order for a single apple to come into existence, earth, air, water, the sun and the properties of the seed and tree, such as germination, growth, photosynthesis and fruiting, must all co-operate. This means that the existence of a single fruit depends on the co-operation of the whole universe. The order and constancy in whatever takes place in the universe is the origin of what science calls "natural laws." This is the name that science gives to the

elements or properties of order and constancy in the universe on which science is based. Yet this has only a nominal existence. What science calls laws may well be the works or activities of God through the agency of angels—through the agency of angels because the Dignity and Grandeur of Divinity requires the agency of some beings like angels or of causes so that people do not attribute directly to God what is disagreeable to them and accuse the Almighty thereof. Because of its obdurate refusal not to make space for religion in its procedures, science attributes the miraculous, purposeful creation and existence and the order, harmony and constancy prevailing therein, which manifestly require the existence of an absolute, eternal knowledge, power and will, either to blind, unconscious, ignorant and inanimate "laws" with only nominal, not real, existence, or to nature itself, which is actually a passive recipient, not an active agent, an object not a subject, and devoid of consciousness, knowledge and will. Or else science attempts to explain existence and life with notions such as chance and necessity. The reason for this compound ignorance of science is that it regards religion as a set of dogmas requiring blind belief and therefore unscientific or irreconcilable with itself. This unforgivable attitude of science and its denial of the existence of the supra-natural dimension of creation or its agnosticism are the result of separating science and religion.

Science Separated From Religion

When separated from religion, science loses its real identity and aim. The aim of science is to study existence in the light of Divine guidance to understand it, to use the universe as a series of ladders to reach the "heaven" of belief and, disposing things in accordance with that belief, improve the world, thereby helping humans fulfill our function as Divine vicegerency on the earth. This is what the Qur'an teaches.

As we read in the Qur'an (2:30–31), when God told the angels that He would appoint a vicegerent on the earth, since vicegerency requires will, knowledge and power, the angels inferred that he would cause corruption, sedition, and bloodshed, and responded: "We glorify You with Your praise and proclaim Your Holiness". The Almighty answered them: "I know what you know not." He instructed us in the "names," that is, the names and reality of things and therefore the keys of knowledge and mastery. As He made us superior to angels through knowledge of things, He regarded "scientific" studies to understand creation and fulfill the role of vicegerent as equal to the glorification and praise of angels. This means that scientific studies done for the sake of understanding creation and thereby recognizing the Creator and improving the world, establishing peace and justice there, are acts of worship. So Islam gives to science and scientific studies a sacred meaning and religious dimension.

Also, the revelation of the Qur'an began with the command "Read!" Having come at a time when there was yet nothing written, this command is significant. This order continues, in the name of the Lord Who creates (Al-Alaq 96:1), which means that humans should study creation in the name of the Lord. The original of the word translated as the Lord is *Rabb*, which means One Who brings up, educates, trains, sustains and raises. This signifies that creation is under the "Lordship" of God and we should study it with all its aspects of birth, growth, and functioning. This is what science does.

A second important connotation of the first revelation is that, as mentioned above, we should study creation in the name of God, that is, to please Him in accordance with the rules He has established. This means that any scientific study should not be contrary to religious and moral injunctions, and therefore should not be made in a way that harms people and changes creation and the order of the universe. When made to discover the Divine "laws" in nature and dispose of it within the limits of Divine permission, any scientific study will not cause environmental pollution, the death of millions of people and destruction of cities; in short, corruption on the earth.

Islam never prevents scientific studies; what it does is appoint for science moral aims and put moral restrictions on it. It aims to urge scientific studies to be for the benefit of humankind as well as other creatures and, by ordering them to be done in the name of God, it raises them to the rank of acts of worship.

When science is separated from religion, although it has brought wealth and material well-being to a very small minority in the world, as especially the last two centuries have witnessed, it can cause world-wide insecurity, unhappiness and unease brought about by scientific materialism, a brutal oppression and colonialism, wide rifts between rich and poor, unending global or regional wars during which millions die or are left homeless or orphaned or widowed, merciless rivalry among the classes of people and dangerous levels of environmental pollution. The separation of science and religion has brought us nothing but great disasters.

Prophets Were Also Masters and Forerunners with Respect to Scientific Discoveries and Progress

It is more evidence of the inseparability of science and religion, even if secular science does not admit it, that Prophets were forerunners of scientific discoveries and humankind's material progress. For example, some interpreters of the Qur'an infer from the verse "When Our command was issued and the oven boiled..." (11:40), that the Ark the Prophet Noah constructed through God's guidance was a steamship. Sailors regard the Prophet Noah, peace be upon him, as their first teacher or patron. Similarly, the Prophet Joseph was the first to make a clock and therefore was considered as the first teacher of clock-makers, and the Prophet Enoch as that of tailors.

However, secular or materialistic science does not regard Divine Revelation as a source of knowledge, or regard revealed knowledge as scientific. For example, it considers the Flood, mentioned in all the Divine Scriptures and oral histories of all peoples, as a myth. If this event cannot be established through "scientific" methods, it will not be scientific and those who regard the methods of science as the only reliable methods to arrive at truth will continue to approach the Divine Scriptures with doubts. This amounts to the denial of Divine Revelation and God-revealed religions. Also, this will also cause many historical facts and events to remain veiled. Furthermore, it is impossible to study and teach correctly the history of the Middle East—Arabia, Egypt, Palestine, Syria, Iraq and Iran, etc.—without considering the life-histories of the Prophets mentioned in the Qur'an. Despite this, by not admitting the "scientific" reliability of Divine Revelation, secular or materialistic science causes many truths to be taught as though they were falsehoods and many false-hoods to be presented as truths, and many realities to remain veiled. For example, the Assyrians who lived in Iraq are presented as having been a pagan people, but we read in the Qur'an ("And We sent him to a hundred thousand or more"—37:147) that more than one hundred thousand peo-ple believed in the Prophet Jonah, who, according to the account of the Bible, lived in Nineveh, the capital of the Assyrians.

Separating science and religion and assigning to each a different realm of competence or relevance is responsible for religion being seen as a set of myths and dogmas—blind beliefs—and science remaining in the darkness of materialism. So, as it is absolutely necessary to "wed" and harmonize mind and heart or the intellect and spirit, it is also of vital importance to harmonize science and religion.

Causality and the Qur'anic World-View

■ Yamina Mermer

T he universe has been made in the form of a book, intelligible, so as to make known its Author. The book addresses humans. The aim is to make us read the book and its parts, and respond with worship and thanks to the will of the Author. We attain to that worship by uncovering, through scientific study, the order in the book of the universe, and displaying the functioning of beings and the workings of the universe.

The universe is not passive. It is not neutral. We cannot interpret it as we wish. There is only one correct way of looking at the world, one universal world-view which is common to all humanity. This view is taught to us in the Qur'an as well as in the book of the universe by our Creator. This does not mean that the Qur'anic world-view does not recognize that the perception of the world differs from one person to another. It allows for plurality within unity so that a universal dialogue is possible. In this world-view there is no fragmentation and no conflict. There is only harmony, assistance, peace and compassion.

The materialist scientific world-view is based on radical fragmentation. Materialist science takes nature to mean a mechanism with no inherent value and meaning. It isolates an object by cutting off its connections with the rest of the world and studies it within its immediate environment.

Whereas our perception of ourselves tells us that we are meaningful and part of the whole universe, and everything must have a meaning and must be part of the whole universe, materialist science has abandoned this. With science taking over, people feel that they have no place in this world. We are isolated from each other. Our lives have no meaning, except in a very limited, egoistic sense. We are alienated from our environment and from ourselves.

In light of modern physics, the mechanistic view is an incoherent description of nature. The developments of modem physics call for a radi-

cal revision of our concept of reality. These developments have shattered all the principal concepts of classical physics. Many concepts, like the causal nature of physical phenomena and the ideal of an objective description of nature, changed with the advent of the new theories of modern science, quantum, relativity and, more recently, chaos theory.

However, these changes have not been matched by parallel changes in the world-view of science. The modifications took place only on a mathematical level, for only the development of mathematical formulations of the behavior of physical phenomena is all that counts for scientists. Such a goal is not regarded merely for its technical utility; rather most scientists believe that prediction of this kind is all that knowledge is about. They claim that our concept of reality is of little or no importance. However, it is clear that our concept of reality has a tremendous effect on how we behave in relation to nature and to other people, and also on the meaning life has for us as individuals. We cannot dispense with a world-view.

This attitude of scientists is in contradiction with modern science. Classically it was thought that science could describe and explain everything in the world "objectively," i.e. as it actually is in reality, and that the "observer," i.e. the scientist himself, could describe the world by means of mathematical models which were independent of his judgment. The discoveries of modern physics, however, point towards the unity of all things, an unbroken wholeness which denies the classical fragmentation of the world into separate and independent parts. In quantum theory, every particle is linked to the rest of the universe and cannot be isolated from it. This oneness of the universe includes human beings as well. Along with abolishing the notion of fundamentally separate objects, quantum theory has introduced the concept of "participator" to replace that of the neutral observer. Modern science therefore restores humans to our central position. It puts an end to the notion of a neutral, objective description of nature and thus to impartial objective science.

Up to the present, materialist science has been based on a deterministic, causal view of the world. Although the latest theories, for example quantum and chaos theories, are leading to a world-view where there is no room for fragmentation and determinism, materialist scientists still insist on following the fragmented and causal approach. They have to be reductionist because they believe in causality. At the same time they do realize that their materialist world-view is collapsing. Theoretically they understand that in order to explain one thing they need to know its connections to all other things. This is obviously impossible because these connections extend in time and in space beyond human capacities; they are infinite and cannot be embraced by human beings, who are also parts of those connections.

The materialist scientists understand that the unity of the universe points to an Absolute Creator. For the things we study do not bear meanings limited to themselves but testify to the Absoluteness of their Creator. But in order to be able to claim that their scientific studies produce knowledge, scientists insist on denying the Absolute Creator. And because their scientific method is based on causality, which cannot accommodate the unity of the universe, they ignore that unity and compartmentalize the universe so that they can study each compartment as the product of a limited number of causes. In this way, they can pretend the universe has no Creator and its meaning is limited to what they tell us about it. They thus claim their science to be the source of knowledge.

There have been many controversies over the conceptual foundations of modem physics. The mechanistic model of reality is not appropriate to modern science. Scientists avoid this issue by adopting the attitude that the paradoxes and contradictions of their science are inherent in nature, thus implying that those paradoxes and contradictions have nothing to do with the inadequacy of their world-view.

But how one can apply the reductionist scientific reasoning to the inseparable universe? It has been widely discussed by scientists and philosophers, but what hasn't yet been realized is that the nature of the materialistic approach to scientific reasoning is incompatible with the unity of the universe. Therefore, either that approach to scientific reasoning or the concept of the unity of the universe has to be reconsidered. That the universe is an inseparable whole is not in dispute. Indeed, the unity observed in the totality of the universe, including us, is so manifest that no one can deny it.

Therefore the materialistic approach to the scientific method has to be reconsidered. This method is reductionist. It reduces everything to fragments then attributes each fragment to causes. However, in reality, all things are interconnected and interdependent. For this reason it is impossible to attribute anything, however small it is, to causes which are themselves transient and contingent. Whatever is responsible for one thing must be responsible for everything. We cannot have one thing without the whole.

Why can we ascribe a thing to its antecedents in time, but not to its neighbor in space? Why should a thing be able to produce another thing just because it happened before? All modern scientists know that space and time are fully equivalent. They are unified into a four-dimensional continuum in which "here" and "there," "before" and "after" are relative. In this four-dimensional space the temporal sequence is converted into a simultaneous co-existence, the side by side existence of all things. Thus causality appears to be an idea which is limited to a prejudiced experience of the world.

Causality leads to the vicious chain of cause and effect. For each cause is also an effect, and the effect is totally different from the cause. Things and effects are usually so full of art and beneficial purposes that their simple immediate causes are lost in a tangle; and even if all causes gathered they would be unable to produce one single thing.

In short, in order for a cause to produce an effect it has to be able to produce the whole universe in which that effect takes place, for that effect cannot exist without the whole universe. They cannot exist separately. Causality is therefore the antithesis of "There is no deity but God," the core of the Qur'anic world-view. Materialist scientists imagine powerless, dependent and ignorant causes to be responsible for the existence of beings and things and thus fancy them to possess absolute qualities. In this way they are implying that each of those causes possesses qualities which can only be attributed to God.

However, the latest discoveries of modern science, like the unity of the universe and the inseparability of its parts, exclude the explanations put forward by materialistic science. They demonstrate that all entities, whether in nature or in the laws and causes attributed to them, are all devoid of power and knowledge. They are contingent, transient and dependent beings. But the properties attributed to any of such entities need

infinite qualities like absolute power and knowledge.

This shows that causality is by no means necessarily to be linked with "objective" study and "neutral" scientific investigation. It is no more than a personal opinion. Moreover, it is an opinion that is irrational, a non-sense. Nevertheless, there is still a widespread conviction that science can do without a Creator. This seemed possible in classical physics, but in quantum mechanics the situation is untenable. Physics is full of examples of ingenuity and subtlety that exclude the causal interpretation and make known the All-Powerful, All-Knowing One. A few illustrations will, I hope, suffice to convince us that the universe with all that is within it is His product.

The idea that the universe began with a Big Bang is something of a paradox. Of the four forces of nature, only gravity acts systematically on a cosmic scale, and in our experience gravity is attractive, a pulling force. But the explosion which marked the creation of the universe required a very powerful pushing force to set the cosmos on its still continuing path of expansion. It is puzzling that the expanding universe is dominated by the force of gravity which is contracting, not expanding. Careful measurements show that the rate of expansion has been fine-tuned to fall on this narrow line between two catastrophes: A little slower, and the cosmos would collapse, a little faster and the cosmic material would have long ago completely dispersed. The materialist scientists can see that such a precisely calculated explosion requires infinite power and knowledge, and yet deny the act of creation. Therefore they are compelled to say, "It just happened; it must be accepted as a special initial condition."

The initial condition, however, had to be very special indeed. And the rate of expansion is only one of countless cosmic miracles. But in their ignorance, they fancy those miracles of Absolute Power to be "remarkable" coincidences, and imply that the universe is a random accident. So, strikingly, the most fundamental theory of recent modern science is totally compatible with the notion of the Absolute Creator. Moreover, it is not compatible with causality. Thus, the need for God, the Causer of causes, enters science in a fundamental way.

Classically, it is believed that a measurement performed in one place cannot instantaneously affect a particle in another distant place. The basis for this belief is that interactions between systems tend to decline with distance. For according to causality a cause has to be in the immediate vicinity of its effect. Otherwise how can particles several meters, let alone light years apart, influence each other's position and motion?

Quantum mechanics, by contrast, predicts a greater degree of correlation, as though the two particles are co-operating by telepathy. This forces us to ask how it is possible to explain this remarkable degree of co-operation between different parts of the universe that have never been in communication with each other without mentioning their Creator. How can they achieve this miracle? Divine Unity is obviously the only reasonable, consistent and acceptable explanation of this miracle and indeed of the universe and all that is in it, including humans.

To the materialists this situation is a paradox because it cannot be explained with causality. But to the believer in God, this is a beautiful aspect of His Unity. It envisions a universal coherence and points to all-encompassing principles that run throughout the cosmos. When we break the vicious chain of cause and effect, the meaningless world of materialism gives way to a world illumined with meaning and purpose. The universe becomes like a vast book addressing us and making known its Author so that its readers take lessons and constantly increase in knowledge of their Maker and strengthen their belief and certainty in the fundamentals of faith.

In short, everything is full of art and is being constantly renewed, and, like the effect, the cause of each thing is also created. For each thing to exist there is need for infinite power and knowledge. Thus there must exist a Possessor of Absolute Power and Knowledge who directly creates the cause and the effect together, which together demonstrate the attributes of their Maker. They are proclaiming the Divine Power and perfection through their own powerlessness and deficiency. They are all announcing, "There is no deity but God."

Just as the universe points to this truth of Divine Unity so does the Owner of the universe teach us this truth in the sacred books He has revealed. The phrase "There is no deity but God" is the fundamental of revelation and it is confirmed by the testimony of beings. It is the key to the Qur'an, which makes it possible to know the riddle of the creation of the universe, a riddle that has reduced materialist science and philosophy to impotence. The path of Unity is the path of Revelation. It is the only path that shows us as our own Master and Owner, and causes us to recognize our True Object of Worship who possesses an absolute power that will guarantee all our needs.

The Qur'an is the only source that teaches us that the universe and the beings within it do not bear a meaning limited to themselves but testify to their Maker's Unity. It teaches what the universe is and what duties it is performing.

For this reason, every Muslim should study the universe and see that all beings, through their order, mutual relationship and duties, utterly refute the false claims of materialist and atheistic reasoning. They affirm that they are nothing but the property and creatures of a Single Creator. Each rejects the false notions of chance and causality. Each ascribes all other beings to its own Creator. Each is a proof that the Creator has no partners. Indeed, when the Creator's Unity is known and understood correctly, it becomes clear that there is nothing to necessitate that causes should possess any power. So, they cannot be partners with the Creator. It is impossible for them to be so. Then the Muslim scientist will say through his investigations and discoveries, "There is no deity but God, alone and without peer."

The universe is a document to be used by the believers. Believing in God is, as the Qur'an informs us, to assent with one's heart to the Creator with all His attributes supported by the testimony of the whole universe. The true affirmation of God's Unity is a judgment, a confirmation, an assent and acceptance that can find its Owner present with all things. It sees in all things a path leading to its Owner. It does not regard anything as an obstacle to His presence. For otherwise it would be necessary to tear and cast aside the universe in order to find Him. and that is impossible for us.

The universe is not the property of materialistic science which has used the universe in a destructive way precisely because it has been unable to find the meaning of the universe. There is no dichotomy between true science and revelation. Rather true progress and happiness for humankind can only be achieved in the way of the Qur'an. All scientific and technological advances are merely the uncovering of the way the universe is created. When the universe is seen to be a vast and meaningful unified book describing its Author and the beings in it as signs of their Creator, these discoveries and advances reinforce belief rather than causing doubt and bewilderment.

The most serious disease afflicting modern humans in our search for happiness and the meaning of life is to regard science, the study of the created world, as separate and irreconcilable with revelation, the word of the Creator. But as we learn to heed the universe and our senses, rather than the materialist scientists, we will wake up to the contradictions of modern scientific reasoning, which more and more people are beginning to realize is no longer valid. Faced with beauty, mystery and purpose, attempts to explain creation with causality are becoming increasingly untenable. Then they will feel the need and importance for true science and knowledge which yield knowledge and belief in God.

The Universe in the Light of Modern Physics

■ Salih Adem

"The least understood aspect of the universe is its being understandable." Einstein

T hese words attempt to pierce the veil of habit that develops in our minds from not looking into the reason for things. The perfection of the order operative in the universe is of such a degree that it prevents us from being aware of it. In the same way, we only become aware of the faultless operation of the watches we have worn on our wrists for years when they stop working.

In the world-view developed upon the foundation of Newton's laws of motion, the universe was likened to a flawlessly operating watch. Events were tied to one another in a cause-effect relationship and our knowing the laws of this relationship allowed us to predict events with great accuracy. It was possible to determine with mathematical exactness a wide range of phenomena, from the times of eclipses of sun and moon to the amount of fuel and the speed needed to put an object into orbit around the earth. The success of these natural laws led many people to believe that they completely expressed and ruled the whole order of the universe.

Because God creates and sustains all things and events from behind the veil of universal general laws, because certain events (causes) are followed reliably by similar events (effects) each time they (the causes) occur, we begin to suppose that the causes are responsible for or create the effects. This is, of course, a gross error, as no number of causes suffices to create even a little effect; for every event, even the tiniest, the whole universe must be presupposed first, including the laws operative within it. Moment by moment, all things and all events are created and sustained by God, Who wills from an infinite range of alternative possibilities a particular actuality.

The clockwork model of the universe derived from Newtonian or classical physics is not a complete account of the phenomena which we observe

112

in the universe. Already in the late nineteenth century, scientists had been bewildered by the lines that turned up in the light spectra emitted by heated gases: the steady, stable clockwork model predicted did not happen. Also, there were problems explaining the behavior of light: sometimes it made more sense as a beam of particles, sometimes as a wave.

Today our understanding of the universe is very far from the clockwork model. The shift in understanding occurred in the first quarter of the twentieth century, beginning in 1900 with the publication of Max Planck's work on radiation. The problem Planck worked on for six years was that the actually measured radiation from hot bodies did not conform to the values predicted by classical theory. He put forward the suggestion that bodies radiating energy did so not evenly and continuously, but unevenly and discontinuously in tiny packets or quanta. So startling was this suggestion that, despite confirmation by experiment, Planck himself thought of his theory as solving the problem of radiation by a sort of trick.

But then in 1905, Albert Einstein published an article using the notion of packets of energy of definite sizes to explain how electrons are ejected from metal when light (radiation) falls on it. Whereas classical theory had predicted that the voltage (measure of the energy of the electrons ejected) would be proportional to the intensity of the light (radiation), Einstein showed that it was proportional instead to the frequency of the radiation. The conformity of this explanation with experimentally observed results gained Einstein the Nobel Prize. (Einstein did not receive the prize for his famous theory of relativity.) The significance of these findings and theories was not fully appreciated at the time.

A few years later in 1910, Ernest Rutherford did a ground-breaking experiment. He bombarded a thin layer made up of gold atoms with high energy particles and showed that the atom contained an extremely small positively-charged nucleus with negatively-charged electrons moving around it. Following the classical model, these electrons should have been small particles orbiting the nucleus in the same way as the planets orbit the sun, steadily losing energy until they fell into the nucleus; in other words, the atom should have been unstable. Again it was a rejection of the classical model, three years later, by Niels Bohr, that helped solve the problem. Bohr argued that the electrons must move in fixed orbits until deflected by the absorption or emission of a unit of energy.

Atoms emit radiation after various external signals and only at specific wave lengths. As Einstein said, every different color of light is composed of energy packets inversely proportional to its wave-length (frequency). Because the Planck constant (h) is very small, the energy of these packets is also very, very small. For example, a normal light bulb emits 1020 light packets (photons) a second. Each of these photons is created when an activated atom or molecule passes to its normal or basic state. Thus light, which allows us to see and which is a basic building block of life, develops as a result of the motions (in wave form) of electrons. The concepts of classical physics could successfully explain many of the events of daily life, but it could not explain events at the subatomic level.

During those years (1910–1925) physics fell into a state of confusion because of the many measurements that conflicted with general theory and could not be explained by it. This situation was to lead W. Pauli (later to discover the principle fundamental to the understanding of the structure and characteristics of elements) to say he would rather have been a singer or gambler than a physicist. In order to explain the observations being made, our perception and understanding required fundamental revision by wholly new methods. This was achieved by Werner Heisenberg, a 24 year-old physicist described by his teachers as a person who dealt with the essence of a subject rather than getting bogged down in detail, a person with powerful concentration and ambition. Perhaps the success of this young mind can be explained by the critical perspective he developed through reading the works of great men such as Kant and Plato, which was later supported with the sound knowledge he got from great physicists. Heisenberg, who relaxed and decompressed from work by climbing rocks and reading poetry, said: "It was around three in the morning when the calculations were completed and the solution to the problem appeared in front of me. First I experienced a great shock. I was so excited that I did not even think about sleeping. I left the house and, sitting on a rock, I waited for the sunrise."

Like the other scientists who established quantum physics, Heisenberg was a philosopher-physicist. The philosophy he accepted and advocated that allowed him to interpret atomic events is as follows: "Even

though it is successful with classical physics, the language we use to explain physical events in the atom or its surroundings is insufficient. For this reason, after making a specific measurement in a quantum system (for example, an atom), using that knowledge we can get a theory that will tell us what kind of results we can find in the next measurement. But it is not possible to say anything about what takes place between the two measurements."

What pushed Heisenberg to make such a statement was that the mathematical tools he used to develop a theory that could explain the observed discontinuity of energy in light and atoms were abstract concepts that had not been used before. In classical physics known numbers are used to give value to matter's position, speed, size, etc. In Heisenberg's quantum mechanics, these sizes are expressed with n x n matrices of infinite dimension, which enabled physicists to calculate the properties attributed to electrons (energy, position, momentum, angular momentum) in an approximate way. Because these abstract mathematical expressions did not have an equivalent in everyday spoken language, it was not possible to approach them with a classical understanding. It was observed that in order to measure the position of an electron, the experimenter necessarily altered its velocity. This problem was formally expressed in 1927 in Heisenberg's famous Uncertainty Principle.

Independently of Heisenberg, Erwin Schrodinger made another significant breakthrough in his mathematical description of electrons. Inspired by the hypothesis put forward two years earlier by De Broglie about the wave properties of matter particles, Schrodinger developed a "wave mechanics" model by which the movement of particles could be calculated. But the fundamental question remained as to what these strange and original waves of matter particles or waves accompanying matter particles were.

The mathematical formulations devised by Heisenberg and Schrodinger are complementary in the sense that physicists use whichever best resolves the particular calculations they are trying to make. There is no formally distinct space between the scientists and the phenomena they are seeking to understand and manipulate: their means of observation and manipulation (the mathematics) in some sense defines the very phenomena whose place (among other properties) they are trying to determine. Alongside the notion of an infinite array of rows and points, as invented by Heisenberg, to plot the position or motion of a sub-atomic particle, physicists and philosophers of physics have begun to speak of arrays of events or stories to try to explain, in something resembling ordinary language, the ideas they are handling. This cannot be described as a world-view in the way that the Newtonian physics confirmed and sustained a world-view, but it is nevertheless a clear and distinct disposition which, instead of excluding God as the Force Who wound up the clockwork and then retired from His creation, admits the in-completeness and uncertainty of human knowledge as a structural element of reality; in other words, the uncertainty is not a function of our present ignorance (to be relieved by future knowledge), but an actual constituent of the way reality is.

Quantum physics, at least figuratively and metaphorically, has become a vehicle for the interpretation of such concepts as matter, beyond-matter, energy, existence and non-existence in a way nearer to Divine sources; and has led to many physicists settling accounts with their conscience and turning towards God Who is understood to be simultaneously transcendent and immanent, there and here.

The Sub-Atomic World and Creation

■ Dr. Şenol Ersin

O ut of the three famous papers that Albert Einstein published in 1905, *On a Heuristic Point of View Concerning the Production and Transformation of Light* explicitly stated the quantum hypothesis for electromagnetic radiation, and *On the Movement of Small Particles Suspended in Stationary Liquids Required by the Molecular-Kinetic Theory of Heat* developed the theory that led to the establishment of the sub-atomic nature of matter.

Following the classical Newtonian physics and under the spell of developments in science, physicists of the nineteenth century claimed that they could explain every phenomenon in the universe. E. Dubois Reymond, at a meeting held in memory of Leibniz in the Prussian Academy in 1880, was a bit humbler: "There have remained eight enigmas in the universe, three of which we are unable to solve yet: The essential nature of matter and force, the essence and origin of movement and the nature of consciousness. The three of the rest that we can solve although with great difficulties are: The origin of life, the order in the universe and the apparent purpose for it and the origin of thought and language. As for the seventh, we can say nothing about it. It is the individual free will." (quoted in Adnan Adıvar, *İlim ve Din* (Science and Religion), İstanbul 1980, p. 282).

The sub-atomic world threw all scientists into confusion. This world and the "quantum cosmology" which it introduces, rather than being a heap or assemblage of concrete things, is made up of five elements: the mass of the electron in the field where an action occurs (M), the mass of the proton (m), the electrical charge which these two elements carry, the energy quanta (h), or the amount of energy remaining during an interaction, and the unchanging speed of light (c). These five elements of the universe can even be reduced to action or energy waves travelling through space in tiny packets, or quanta. Since the quanta required for an action are unique to it and exist independently of the quanta required for the

previous action, it becomes impossible to predict the exact state of the universe. If the universe is in t1 state now, it cannot be predicted that it will be in the same in t2 state later. Paul Renteln, assistant professor of physics at California State University, writes:

> "Modern physicists live in two different worlds. In one world we can predict the future position and momentum of a particle if we know its present position and momentum. This is the world of classical physics, including the physics described by Einstein's theory of gravity, the general theory of relativity. In the second world it is impossible to predict the exact position and momentum of a particle. This is the probabilistic, subatomic world of quantum mechanics. General relativity and quantum mechanics are the two great pillars that form the foundation of twentieth-century physics, and yet their precepts assume two different kinds of universe."[48]

The real nature of this sub-atomic world and the events taking place in it make it impossible to construct a theory to describe them, because they cannot be observed. One reason for our inability to observe them is that, as Renteln writes in an attempt to propose a theory which he calls quantum gravity to reconcile the two different worlds of classical and quantum

---

48   *American Scientist*, Nov.-Dec, 1991, p.508.

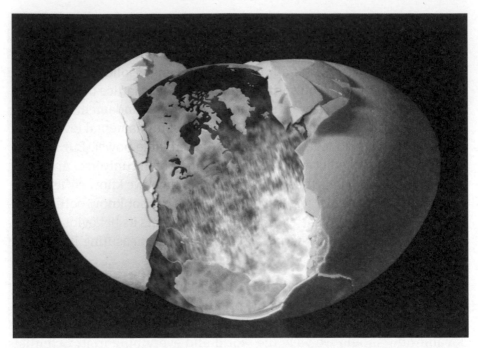

physics, 'the events take place at a scale far smaller than any realm yet ex-plored by experimental physics. It is only when particles approach to with-in about ten to thirty-five meters that their gravitational interactions have to be described in the same quantum-mechanical terms that we adopt to understand the other forces of nature. This distance is 1024 times smaller than the diameter of an atom, which means that the characteristic scale of quantum gravity bears the same relation to the size of an atom as an atom bears to the size of the solar system. To probe such small distances would require a particle accelerator 1015 times more powerful than the proposed Superconducting Supercollider.'

At the outset of this century, electrons surrounding the nucleus of an atom were thought to orbit the nucleus like planets in a miniature solar system. However, later research has modified that view. The electron is now understood to be more of an energy field cloud fluctuating around a nucleus. The nucleus itself seems to be composed of two smaller con-stituents—protons and neutrons. However, in the 1960s, physicists Mur-ray Gell-Mann and George Zweig confirmed by experiment that protons and neutrons were made up of even more elementary particles, which Gell-Mann called 'quarks.'

Quarks cannot be seen, not just because they are too small but also because they do not seem to be quite 'all there.' Quarks are better described as swirls of dynamic energy, which means that solid matter is not, at its fundamental level, solid at all. Anything seemingly solid thing you hold in your hand is really a quivering, shimmering, lacy lattice of energy, pulsating millions of times every second as billions of fundamental particles gyrate and spin in an eternal dance. At its most fundamental level, everything is energy held together by forces of incredible power. But not only are we unable to predict even the nearest future of the universe. According to Werner Heisenberg's theories, at any given time, we know either where a particle is or how fast it is travelling, but we cannot know both. This is because the very act of measuring the particle alters its behavior. Measuring the particle's speed changes its position, and measuring its position changes its speed.

However, the unpredictability in the sub-atomic world does not change anything in our everyday, predictable world. Everything works according to the basic laws of classical Newtonian physics. Why is this so and how should we view the world and everything in it? Scientists who believe in the existence of God and His creation of the universe suggest that creation was not a single event. That is, God did not create the universe as a single act and then leave it to operate according to the laws He established. Rather, creation is a continuous act (creatio continua). In other words, roughly like the movement of energy or electricity and its illuminating our world by means of bulbs, existence continuously comes from God and returns to and perishes in Him. Through the manifestation of all His Names, God continuously creates, annihilates and re-creates the universe. Some medieval Muslim scholarly saints such as Muhy ad-Din ibn al- 'Arabi and Mawlana Jalal al-Din al-Rumi called these pairs of acts as the continuous cycle of coming into existence and dying. Because of the incredible speed of this movement, the universe appears to be uniform and continuous. Rumi likens this to the fast spinning of a staff on one end of which there is fixed a light. When spun at speed, the light on the end of the staff appears as if a circle of light. Unable to explain the extreme complexity of existence and the events taking place, some scientists assert that everything is in chaos and attribute the formation of the universe as it is to mere chance. According to them, other universes could have formed, yet sim-

ply did not, and there is no reason that the universe is the way it is. Given that it is impossible for even three or more unconscious things moving at random to come together by themselves to form even the simplest entity, it is highly questionable whether a rational person can accept that the wonderful order prevailing in the universe can be explained without attributing it to a supernatural intellect. A. Cressy Morrison writes:

The proverbial penny may turn up heads ten times running and the chance of an eleventh is not expected but is still one in two, but the chance of a run of ten heads is very small. Suppose you have a bag containing one hundred marbles, ninety-nine black and one white. Shake the bag and let out one. The chance that the first marble out is the white one is exactly one in one hundred. Now put the marbles back and start over again. The chance of the white coming out is still one in a hundred, but the chance of the white coming out first twice in succession is one in ten thousand. Now try a third time, and the chance of the white coming out three times in succession is one hundred times ten thousand or one in a billion. Try another time or two and the figures become astronomical.

The results of chance are as closely bound by law as the fact that two and two make four. All the nearly exact requirements of life could not be brought about on one planet at one time by chance. The size of the earth, the distance from the sun, the thickness of the earth's crust, the quantity of water, the amount of carbon dioxide, the volume of nitrogen, the emergence of man and his survival-all point to order out of chaos, to design and purpose, and to the fact that, according to the inexorable laws of mathematics, all these could not occur by chance simultaneously on one planet once in a billion times. It could so occur, but it did not so occur.[49]

49   Morrison, Cressy. *Man Does Not Stand Alone*, New York, pp.98-9.

Attributing the impossible to chance is a trick of the human mind, its stubborn resistance, which confuses a theoretical possibility with the actual facts. For example, it is possible that the Pacific Ocean has now changed into milk, but actually it has not. As it is impossible to construct a building on a flowing stream, God Almighty spread over the unpredictability of the sub-atomic world the veil of the speed of its movement and made the universe dependent on what we call laws. It is for this reason that everything in the outer face of nature works according to the basic laws of classical Newtonian physics. However, it is a matter of controversy between the two schools of Ahl al-Sunna wa 'l-Jama'a whether the universe has a continuous existence working according to established laws, and things accordingly have perpetual properties; or whether God continuously creates the universe and orders the actions and behavior of each component at every moment. The followers of the Maturidi School assert that God created the universe and set it to operate according to certain laws which He established, giving each thing certain properties. For example, fire burns because God gave it the quality of burning. But the followers of the Ash'ari School maintain that the universe does not have a perpetual, established existence and reality. Nor do things have essential qualities of themselves. God creates the universe anew each 'moment' and directs it continuously by ordering each thing to do what it must do. For example, fire does not essentially have the quality of burning; rather, God gives it the order to burn and it burns. Since according to the dictates of life in the universe, He usually orders it to burn, we think that fire essentially has the quality of burning.

As we accept the 'relative' truth of both Newtonian and quantum physics at the same time, we can also accept the truth of the views of both schools of Muslim faith. As illustrated by belief and by life at the most fundamental level of existence, the sub-atomic world, God is continuously active, creating the universe anew and directly administering it. Yet at practical level, life is impossible for us if we do not accept or assume the uniform continuity or stability of existence. What would life be if we were conscious that the sun would not rise tomorrow morning or that we might not live a second longer? It is theoretically conceivable both that the sun might not rise tomorrow and that we might not survive a second longer, although highly unlikely.

Quest to Solve the Mystery of Life

■ Fatih Kocabaş

The quest to solve the mystery of life seems to be continuing. Where did we come from? What's matter and what's beyond it? Where and how did life originate? What about the Adam and Eve of other organisms? Obviously, we were not allowed to witness either the creation of universe or the beginning of life on Earth. We don't know much about creation, but we can see the results of creation.

While discussions on the education of creation in schools continue, generations grow up with lack of knowledge about the Creator and understanding of His Actions. The current education system in high schools and colleges is giving knowledge about the universe, nature, earth, and life but courses are not directed to help students to understand the Actions of the Creator. How and where can people learn about their Creator? Although there are various means such as the internet, religious groups, and journals, it is not always feasible and sufficient to understand directly the Creator's Actions without a good understanding of sciences. Fortunately, every science continuously mentions God with its unique language and speaks of the Creator, but we will need a point of view, a window to see beyond our sight and knowledge, just as we need a microscope to see microorganisms or a telescope to discover the depths of the universe. With some attention, everyone can understand how sciences reflect God's Actions. That's why we should listen to what sciences tell us in their own language.

Imagine there is a simulation program to analyze a car crash. In this program, let us say you enter different parameters such as velocity, weight, angle of hit, general structure of the car, hardness of the body, and weather information like wind velocity and direction and so on. After you click on the OK button in this imaginary simulation program, you almost get the same results as those obtained by real physical crash tests. This obviously shows us a skillful software programmer and his

great knowledge of mathomatics and physical events. Nature is composed of millions and millions of parameters determining a final result, just like this simulation program. For instance, when you throw a stone into a lake, first it falls with a velocity, and then you see a wave of water expanding into its surrounding from the center. The velocity of this stone at a certain time and place and wave formation on the surface of water can be explained with some physical laws described by mathematical equations. Whoever instituted these rules for physical events also created the universe in a perfect mathematical order. From these and the knowledge we get from computer sciences, physics, and mathematics, we can open windows in order to understand the ruler of the universe as Glorious Creator.

We are at a time of great advancements in gene technology and a huge increase in knowledge about molecular biology; even individual structures of biological molecules are known and many more facts have been

discovered about cellular mechanisms. The more we learn, the more we face complexity and organization in the even the tiniest compartments of a cell. The cell is no longer a small room filled with a gel-like structure, it is a massive factory that contains all required machinery and it is automatic, well balanced, and continuously renewed. Things are in constant motion; uptake follows release of substances and a signal from outside results in a response produced inside. With this increase in understanding of how living things work and with the necessity to discover how these things originated, scientific research is engaging in furious debates. Some scientists like to talk only of scientific platforms or about so-called testable scientific subjects, but this does not change the reality. We wonder about the origin of life, and we wish to live forever. We are finite but dream of infinity. How can we think of eternal life if we were a product of something that is not eternal?

Imagine there is a high-tech but small self-sustaining factory producing highways and trucks to carry items, fuel engines for the energy that can be used in many different processes, and containing solar energy collectors. There are great photocopy machines for the production of a new factory, and feedback systems to control and repair any problems as well. Without any effort, a perfect engineer, scientist, architect, and chemist control all these events and thousands of machines, engines, highways without any problem. Similarly, believe it or not, the cell is an excellent composition of around one million molecules, thousands of machines, and energy-producing engines. There are highways, trucks, feedback systems and more exactly arranged and finely controlled. Mitochondria, for instance, is one of the most essential cellular organelles and produces ATP molecules, energy obtained from organic molecules for cellular processes. In addition, cellular requirements vary by time, and vesicles carry required molecules as cargo on molecular motors using microtubule pathways to different places. Those and many other examples we learn from the biological sciences, and we are thus pointed to the Glorious Creator of the Earth.

When you consider a cell coming into existence by causes other than the hands of a Creator with numerous levels of regulation, the coordination of subcellular structures such as organelles, information storage in DNA, and use of this information required for their specific function, it means molecules come together under the effect of natural causes and form an artistic cellular structure in an intelligent manner. Actually, this reminds us of a very famous experiment by Stanley Miller to make amino acids, the building blocks of proteins, to demonstrate that life on earth originated by natural causes and chance. Miller took molecules which were supposed to represent the major components of early Earth's atmosphere and put them into a closed system. He used methane (CH_4), ammonia (NH_3), hydrogen (H_2), and water (H_2O) in his experiment and ran a continuous electric current to stimulate lightning storms and to drive these unfavorable reactions. He found that three amino acids were synthesized in these conditions. Later, it was found that this composition was different from the early Earth's atmosphere and arguments raised against his experiment, because he assumed continuous energy input not possible in nature. However, this was exciting at that time, as illustrated by headlines like "Miller created life." On the other hand, what Miller had managed to synthesize was only a few inanimate lifeless molecules.

People who do not believe in God also do not believe in creation. That's why they tend to conclude that "nothing is created out of nothing, and nothing goes to nothing; there is only composition and decomposition." But the All-Powerful One has two ways of creating. The first way is through origination and invention and the second way is through composition and through art. He creates from nothing together with everything necessary for, again, nothing. In the second way of creating, He forms beings from materials of the universe in order to show his delicate wisdom, perfection, and the manifestations of His Names. *"O people! be careful of (your duty to) your Lord, Who created you from a single being and created its mate of the same (kind) and spread from these two, many men and women..."* (An-Nisa 4:1)

When we think about the lessons learnt from these examples and also from natural sciences with their special focus areas, we realize that every science somehow declares the glory of the Creator of this universe. However, there may not be an opportunity in school to discuss and go deep into the understanding of the Actions of the Creator. With the window of what sciences open to us about God, we can uncover the hidden truths.

Science and Religion: Between Friction and Harmony

■ Alphonse Dougan

C an science and religion coexist? Can an inquisitive mind adopt any religion? Are faith and scientific inquiry incompatible? Is religion a set of dogmas and hence closed to scientific investigation? Is scientific investigation as objective as claimed? Is reality limited to what science discovers? These and similar questions have occupied philosophers, scientists, and people of faith since the Renaissance. If religion were the "opiate of the masses," we could not expect an inquisitive mind to adopt any religion. But countless critical thinkers and scientists believe in a God that hears and responds to their prayers.[50]

What we mean by religion and science affects how we answer such questions. Therefore we must agree on common definitions. A study of the scientific method, where and how it is applied, is likely to shed light on the perceived conflicts between science and religion.

Science and Scientism

In broad terms, science is a systematic way of exploring the universe. The scientific method helps us discover facts that cannot be directly observed. As described in SGNA:

> "Though we may be unable to observe an aspect of the universe directly, we may deduce its existence and its properties by observing the effect that it has on those phenomena that we can observe. In other words: by explaining the observed aspect of the universe, we go one step beyond that of mere observation, and we gain knowledge about something that we have not observed directly. This is the whole point: we gain information from sources other than direct observation. Use of the scientific method ensures that this information is accurate, and not influenced by the subjective points of view of a single researcher or the use of inaccurate instruments."[51]

---

50   National Institute for Healthcare Research: http://www.nihr.org
51   The Skeptic's Guide to the New Age http://www.euronet.nl/users/frankvw/sgna_5.html

But there is a difference between accepting the scientific method's discoveries and accepting as truth only what science discovers. The latter, which Huston Smith called scientism, is "the belief that no realities save ones that conform to the matrices science works with—space, time, matter/energy, and in the end number—exist."[52]

The successes of science and technology, and their applications, have created a kind of utopia where science, especially positive science, has become the source of all knowledge and wisdom. In the views of positivist philosophers like Hume, Locke and Berkeley, anything that cannot be measured does not exist. But closer examination reveals the oversimplistic nature of this view.

Many contemporary scientific theories talk about subjects that cannot be directly measured. Take the atom. Physics textbooks are full of diagrams depicting it as consisting of a nucleus with and orbiting electrons. The diagrams may be quite sophisticated, and the explanation of how the system works can be quite detailed. Yet nobody has ever seen an atom. The closest we have come is seeing their positions via a Scanning Tunneling Microscope. But this has not prevented us from discovering the details of an atom's inner workings.

---

52   Huston Smith, *Forgotten Truth: The Common Vision of the World's Religions* (San Francisco: Harper, 1993).

The Scientific Method

When we want to discover new information about a subject, we first use direct observation, which has the highest degree of certainty. The scientific method establishes guidelines and procedures for objective, accurate, and systematic observation. The most dependable direct observation is the one that can be repeated and has known parameters. By repeating the observation under the same parameters, other scientists can verify a statement's truthfulness.

When direct observation is not possible or insufficient, the thought process steps in. We infer and deduce based on observation. We hypothesize and look for exceptions. Such verification is where the scientific method really shines: It brings an objective mechanism for testing hypotheses to the discovery process. It helps us decrease the degree of uncertainty regarding that which cannot be observed directly.

Controlled and repeatable experiments are the next best techniques, for they enable us to obtain objective and sound knowledge. While we cannot control, we still can observe and infer. However, our degree of certainty and accurate knowledge decrease as we move further away from direct observation.

The scientific method's main purpose is to decrease such uncertainty and to ensure that it is not affected by individual bias or equipment error. Our level of control while observing a phenomenon determines the level of our knowledge's certainty. While the media or popular culture labels certain statements scientific, this characteristic depends on the nature of the verification process. Some so-called scientific facts are direct observations; others are theories that require a thorough testing.

Not all scientists agree on what constitutes the scientific method. Some describe it as the collection of all means and methods scientists use to investigate a phenomenon. Since this definition is too broad, we will focus on a narrower one accepted by most scientists: The scientific method consists of the following:

1. Defining the problem and making repeated observations to collect information,

2. Forming a hypothesis to explain the observed phenomenon,

3. Testing the hypothesis by matching it against other observations,

4. Developing a theory consistent with your observations,

5. Using it to make predictions,

6. Testing predictions by repeated, preferably controlled, experiments and/or further observations,

7. Modifying the theory as indicated by your results,

8. Repeating steps 5 through 7 as necessary,

9. Reporting the research notes and results for professional review.

These steps can be summarized in three stages: observing, theorizing about underlying causes, and verifying through more observations. Theorizing is the key step. The other steps require hard work and can be done by any competent, knowledgeable worker. Developing a theory, however, requires a flash of insight, sometimes called intuition.

Coming up with new ideas is part of what makes a great scientist. Despite its being the basis from which all scientific work proceeds, we cannot study or explain this scientifically. We may call it a gut feeling, hunch, inspiration, or insight, but we still do not know its source and cannot schedule it. We can encourage and stimulate it, but we cannot control it. The scientific method helps us ensure that what comes out of intuition is sound and objective, but does not tell us how we come up with the idea. So the scientific method really is about verification.

Limitations

The main mechanism of verification is experiment and observation. While a powerful tool, verification is limited by its definition: If we cannot control a phenomenon or make proper observations, we cannot develop an idea into a scientific theory.

How do we establish a proper experimentation environment? First, we must set up a controlled experiment to control all the factors involved, except for the two factors whose relationship we are investigating. This involves a control and an experimental group. The control group is normal (as a basis for comparison), while the experimental group differs from the control in only one area.

We then allow for the experimental variable, defined as the one area of difference between the two groups. If we set up two groups of subjects with only one difference and our observation confirms a correlation between this factor and a result, we can safely say that that factor is a cause of that result. If we cannot establish two identical groups, the next best option is to try to average out the differences by selecting the group's members so that no factor is represented disproportionately. This is usually possible only when working with inanimate objects.

It is extremely difficult to control all involved factors, as well as to conduct repeatable experiments, when the subjects are people. Since all people and societies are unique, it is very hard to repeat any psychological or sociological experiment. Also, observing people often causes behavior modification. Thus, some scientists have debated whether psychology and sociology, among others, are really sciences.

Several essential questions of personal and social life fall into this category of phenomena: "Questions about the origin of thought, about the origin of intuition or about creativity often lead into the realms of philosophy, if not existentialism. What makes the human mind work? Where does sentience come from? What is the 'I' that seems to live three or four inches behind my forehead and thinks it is me? And how are we ever going to apply the scientific method to answer these questions?"[53]

This leaves us with a dilemma: What should we do when confronted with an idea that is hard or impossible to verify scientifically? As noted in SGNA: "In modern science, the 'scientifically correct' approach in that case is usually to reject the idea. As long as we can't prove that the idea is correct, it must be assumed to be incorrect. But that approach ignores the fact that the scientific method cannot be used to answer all questions."[54] Science can tell us almost everything about our body, except the age-old questions: Why? Why am I here? Who am I? What is the universe and why was it created? According to SGNA: "Science has never really dared to tackle these subjects. The questions are labeled 'existentialism' or 'philosophy' and 'appropriately filed.'"[55]

---

53   *The Skeptic's Guide to the New Age.*
54   *Ibid.*
55   *Ibid.*

The Delicate Balance of Life on Earth

■ Dr. Haluk Nurbaki

"We have given the produce of the earth in harmonious balan-ce and proper proportion[56]."

I t is surely not easy, on first reading, to divine the great scientific mes-sage imparted by this verse. The scientific description of the earth con-tained in the verse must be a humiliation to those who see the earth as no more than an accident in the universe. The 23.5-degree inclination of its axis is a matter of such delicate computation that it is impossible for both physics and philosophy to pre-calculate it. For example, if the earth's axis inclined by 25 degrees, the polar caps would melt in a few hundred years, and the oceans would be flooded with ice. If the slant was 22 de-grees, on the other hand, Arctic ice would invade the whole of Europe, and life would be possible only in the equatorial regions of the earth. The All-Mighty emphasizes this important insight at the beginning of the verse in the description of spreading out and ordering.

Again, this spreading out and ordering of the earth is closely related to the rotation on its axis every 24 hours. If the earth had completed one revo-lution every 30 hours, such tremendous winds would have ensured that the earth would have become a hurricane-ridden desert. If, on the con-trary, the earth had rotated every 20 hours, most plants would have been unable to complete their biological activity and fallen victim to droughts.

The spreading out and ordering mentioned in the first part of the sacred verse, then, is obtained for the rotation of the earth by the harmonious in-clination of the earth's axis. If this result were left to chance, it could be achieved only after millions of trials, and probably not even then. God's presentation in various parts of the Qur'an of His marvels of order and measure, aims at closing all doors to unbelief with the mathematical and physical order of the earth and the universe. The most significant message of the verse is the proper balance of things produced from the earth. What are these things, and what are these harmonious measures?

---

56   Al-Qamar 54:49.

Scientific research to date links the chain of life to the balanced interactions between plants, animals and bacteria. The bacteria are charged with transferring nitrogen from animals to plants. Plants produce the oxygen needed by animals and other organisms, and animals supply both carbon dioxide and through bacteria nitrogen to plants. While the chain of life proceeds in this manner, it is imperative that the proportion of oxygen in the air remain close to twenty percent. This is where the most delicate of subtleties begins. All smoke and exhalations are converted by plants into oxygen. One would need a computer to calculate the ratio of plant species needed to maintain the oxygen in the air at twenty percent.

There has to be a divine computer that can regulate the amount of plants needed for the smoke from chimneys and the oxygen consumption of humans and provide the necessary oxygen to the air. This incredible calculation can only be considered as a divine miracle. The Qur'an declares that "the things produced in the earth are subject to proper balance" and its wisdom reaches us across a gulf of fourteen centuries, from a time when none of the above facts were known.

Millions of years ago, a vast blanket of flora covered the earth. The purpose of this was to increase the oxygen balance of the atmosphere. Dinosaurs, the animals benefitting from these plants, roamed on earth. Finally, the oxygen ratio began to exceed twenty percent. The consumption of the plants by dinosaurs and their exhalation of carbon dioxide were no longer able to balance the gigantic production of oxygen by the plants. At this point, a vast geological upheaval occurred, and both the flora and the dinosaurs vanished from the face of the earth. Then, God produced all fish, birds and mammals at the same time. (That is the latest valid hypothesis, the views of the diehard evolutionists having been superseded some time ago.)

As the verse tells us, the volume of plants is in such a harmonious balance that a tree has been assigned to purify the fumes of every smoking chimney. Humans, even those who think they believe, are so heedless that we can never fathom these delicate calculations of Divine Omniscience, and for this reason we cannot intuit the secret of the Provider of the worlds. The respect accorded in Islam to trees and the importance assigned to planting new trees stand in recognition of the above facts. Let me now provide some further incredible calculations.

For nearly every illness, the Great Creator has fashioned a plant or microbe as a cure. How can the ignorant, altogether lacking evidence, call this system a coincidence! To create the earth, to settle human beings there, and then to keep ready all the botanical and bacterial remedies for all their ailments in the laboratories of nature! There are enough foxglove plants in the world to provide digitalis to cure all heart patients. There are also enough hashish plants to ease the sufferings of all the painfully ill, yet the medicine of that plant has become black-market contraband under the pressure of heedless selfishness and is used to provide pleasure for the lunatic fringe.

Another example of the things produced in harmonious balance in the earth: until about a hundred years ago, firewood met our heating and energy requirements. If coal and oil had not been discovered, the forests would all have been the last of their kind on earth. But just at the right moment, the divine computer delivered the coal and later, the oil, that it had prepared millions of years ago, and in such measure as to provide enough for all human beings. Unfortunately humans, prisoners of egotism, are now preparing for the greatest war in history with oil at the center of the controversy. And what of the House of Islam? Because it has not truly embraced the Qur'an, because it has been unable to understand it and to realize new scientific advances, it now stands dazed, looking in bewilderment at even the wealth gushing from its own backyard.

Let us now look at the proper balance of the earth's constitution in terms of its metals: We do not know the proportions of metals in the central core of the earth and the liquid mantle surrounding it. But in the crust on which we live, the elements are distributed in such proportion that it is as if a scientific committee had provided a shopping list and the orders had been met by an infinitely powerful factory. Each substance is present in

the earth's crust to the exact proportion that the level of civilization ordained by God demands. Compounds of silicon, iron and potassium are the basic substances for residential construction. If even one of these had been missing, we could not have cities.

Until recently we did not know what a blessing water is. Today we know that the calcium bicarbonate in water is the best aid to digestion. Vital substances such as salt are distributed over the earth in such proportion that hawse have almost arrived in a fully equipped biological laboratory. Have you ever considered that sea water has been evaporating, and then returning to the seas by rivers, for millions of years? During this process, new substances are transported to the seas from land, and yet the composition of sea water never changes. Observe the magnificence of this miracle of the divine computer: millions of events take place, yet the harmonious balance imposed by God on the earth's produce does not change. For the Guarded Tablet is also, in a sense, a law of the Qur'an.

Returning to metals, there are some whose names have been heard of only in the last 150 years, such as beryllium, uranium, cadmium, tungsten, tantalum and gallium. When these were first discovered, everyone regarded them merely as laboratory curiosities. Only later was it realized that these are the indispensable building blocks of advanced technology. From the utilization of atomic energy to high temperature technological activities, each one of these metals represents some essential property, and their presence on earth is adjusted according to the part they were predestined to play.

One of the greatest wonders of the planet earth, which the All-Mighty furnished before lowering us upon it, are the radioactive substances of the world. These are present in the earth's crust in such a perfect proportion that its measure could not be ordered by any scientific committee. On the one hand, there is uranium-235 to provide nuclear power, safe in its natural setting yet dangerous when purified; on the other hand, carbon-14 to enable biological activity. And how wonderful are mineral springs whose waters, which bear moderate amounts of radioactive substances, dispense health to millions of people all over the world.

Consider what we have said about the radioactivity of the earth from the reverse point of view. If uranium had been present in the earth exclusively in the form of its uranium-235 isotope, the world would have become a witch's cauldron a short time after it was formed. On the other hand, if uranium-235 had not been present in uranium-238 in the proportion of 0.7%, we could not have obtained atomic energy. God has invested uranium-235 with such a characteristic that it can be converted to nuclear power only when it is separated, and does not pose a danger in its natural matrix in uranium-238.

Many biological events could not take place but for the presence of carbon-14 in the atmosphere. If this substance, which has a proportion of one ppm (parts per million), had been slightly more common, it would have constituted a lethal hazard. And if the sodium-24 isotope were present in mineral springs, taking a bath would be like being present at Hiroshima when they dropped the bomb. Although the main substance in mineral springs is sodium, isotopes other than sodium-24 are predominant.

Indeed, we could never exhaust commentary on this verse if we were to fill volumes. I have therefore been content to offer only a short summary. Let us read and re-read it with due reverence and wonder: "*We have given the produce of the earth in harmonious balance and proper proportion.*"

Index